Military and Humanitarian Health Ethics

The interdisciplinary book series Military and Humanitarian Health Ethics fosters an academic dialogue between the well-established disciplines of military ethics on the one hand and medical ethics, humanitarian ethics and public health ethics on the other hand. Military and Humanitarian Health Ethics have emerged as a distinct research area in the last years, triggered among other things by the unfortunate realities of armed conflicts and other situations of humanitarian disasters - man-made or natural. The book series focuses on the increasing amount of ethical challenges while providing medical care before, during, and after armed conflicts and other emergencies. By combining practical first-hand experiences from health care providers in the field with the theoretical analysis of academic experts, such as philosophers and legal scholars, the book series provides a unique insight into an emerging field of research of high topical interest. It is the first series in its field and aims at publishing state-of-the-art research, illustrated and enriched by field reports and ground experiences from health care providers working in armed forces or humanitarian organizations.

We welcome proposals for volumes within the broad scope of this interdisciplinary and international book series, especially proposals for books that cover topics of interest for both the military and the humanitarian community, and which try to foster an exchange between the two often separate communities of military and humanitarian health care providers.

Daniel Messelken • Ana Elisa Barbar
Editors

Challenging Medical Neutrality

Healthcare ethics in armed conflict and other complex settings

Springer

Editors
Daniel Messelken
Center for Ethics
ZH Center for Military Medical Ethics
University of Zurich
Zürich, Switzerland

Ana Elisa Barbar
Insecurity Insight
Geneva, Switzerland

ISSN 2524-5465 ISSN 2524-5473 (electronic)
Military and Humanitarian Health Ethics
ISBN 978-3-031-69397-7 ISBN 978-3-031-69398-4 (eBook)
https://doi.org/10.1007/978-3-031-69398-4

This Springer imprint is published by the registered company Springer Nature Switzerland AG
The registered company address is: Gewerbestrasse 11, 6330 Cham, Switzerland

Contents

About the Contributors

Ana Elisa Barbar is an expert on protection of healthcare, with a focus on safe health responses, civil-military relations, and emergency preparedness. She has worked with the World Health Organization and in the Red Cross and Red Crescent Movement in different roles connected to the well-being of health professionals and protection of healthcare in contexts of conflict and violence. She has been deployed in Latin America, Middle East, and Africa, and worked as global advisor to policy and operations of the International Committee of the Red Cross on issues of protection of healthcare. Ana Elisa is a psychologist with a clinical license and a full residency in Primary Health Care and an Executive Master's in Policy Making and International Negotiations.

Martin C. M. Bricknell was appointed as Professor of Conflict, Health and Military Medicine at King's College London in April 2019. Prior to this, he served 34 years in the UK Defence Medical Services, culminating his service as the Surgeon General of the UK Armed Forces. He undertook operational tours in Afghanistan, Iraq, and the Balkans with multiple additional overseas assignments. In 2010 and 2006, he held senior medical adviser appointments in the NATO ISAF mission. He was awarded the Companion of the Order of Bath, the Order of St John, and the US Bronze Star during his military service. He is an accredited specialist in General Practice, Public Health and Occupational Medicine. His multiple academic papers cover: how organizations learn, care pathways in military healthcare, military healthcare ethics, civil-military relations in health, and the political economy of health in conflict. He is also Deputy Director of the KCL Centre for Military Ethics, Veterans Adviser for the King Edward VII hospital, Editor-in-Chief of the Military Medical Corps Worldwide Almanac, a non-resident Fellow of the Centre for Global Development, and on the editorial boards for the *Journal of Military and Veterans Health* and *BMJ Military Health*.

Innocent Okafor Eze is a medical doctor with over 15 years of clinical experience. He is currently serving as the chief consultant and head of the Department of Obstetrics and Gynecology at a military reference hospital in Southern Nigeria. He was enlisted into the Nigerian Navy as a medical consultant and was commissioned to the rank of lieutenant commander in 2018. Since then, aside professional military duties and experience, he has practiced in busy military tertiary health facilities for over 5 years. This has significantly exposed him to the conflicts and impacts of dual professional loyalty on ethical medical practice by military health professionals. Innocent is naturally attracted to issues on moral philosophy and has keen interest in pursuing a career and further research in bioethics. He has also been actively involved in training interns, house officers, and postgraduate resident doctors. He has actively participated in national and international academic conferences and workshops as a speaker and published some works.

Paul Gilbert was Professor of Philosophy at the University of Hull where he taught for 40 years and was head of department for 17. He has published numerous books and articles mainly on topics of relevance to contemporary social and political issues. Paul Gilbert passed away in 2022.

Erik Gustavsson is Associate Professor in applied ethics with a special focus on medical ethics at Linköping University in Sweden (https://liu.se/en/employee/erigu18). He defended his doctoral thesis on needs in health care priority setting in 2018. Since then, his main research interest relates to medical ethics, especially ethical issues that arise in health care priority setting. During 2023 he primarily works in the project *From scarcity to extreme scarcity: priority setting in crisis and war* funded by The Swedish Civil Contingencies.

Uchenna Clara Innoeze is a clinical psychologist with over 10 years of clinical experience, having worked in both private and public sectors including International Federation for Psychotherapy (IFP) and Police general hospital as a clinical psychologist. She obtained her MSc in clinical Psychology in 2013 and since then has worked with nonprofit organizations, including African Network for the Prevention and Protection against Child Abuse and Neglect (ANPPCAN) Nigerian chapter, Prisoners Rehabilitation and Welfare Action (PRAWA) and Carmelite Prisoners Interest Organisation (CAPIO). She obtained PhD in clinical Psychology in 2020. Her experiences in development work have enabled her to develop passion for the vulnerable in the society, particularly young people and women. Her teaching and clinical experiences availed her the opportunity of publishing papers on mental health and also to collaborate with professionals in other discipline such as obstetrics and gynecology as well as psychiatry. She has equally participated as speaker in both national and international conferences.

Julian W. März, Dr. iur. et Dr. sc. med is a research fellow at the Institute of Biomedical Ethics and History of Medicine (IBME) an affiliate at the Competence Center Medicine-Ethics-Law Helvetiae (MERH) of the University of Zurich. He has studied law, medicine, and bioethics at the Universities of Zurich, Oxford, Munich, Regensburg, and Passau, and the IEP de Paris, and has worked as researcher for the German Cancer Research Center (DKFZ, Heidelberg) and the Centre for Law, Medicine, and Life Sciences of the University of Cambridge. He is currently advising the World Health Organization (WHO) on a number of topics in the area of public health ethics, global health ethics, and digital health ethics.

Duncan McLean has managed operations with Médecins Sans Frontières in both field and HQ beginning 2002. Since 2013, he has collaborated with MSF on a number of dossiers such as critical incident reviews and operational/context analyses. He also contributed chapters to MSF book publications *Saving Lives and Staying Alive* and *The Politics of Fear*. Duncan holds a master's degree in Political Science and a PhD in history. Alongside his MSF-related research, he has worked with International Crisis Group and lectured at several universities in Prague, focusing on epidemiological and colonial history, and conflict studies. He is currently based in Geneva working with MSF-Switzerland as a senior researcher.

Melissa McRae is an Australian Doctor, dual specialized in emergency medicine and public health with an advanced master in bioethics. She has worked in the humanitarian sphere for over 20 years in a range of roles including clinical care, medical operational strategy (development, implement, and evaluation), research, management and leadership, and medical ethics. Melissa has experience working for government, non-government, and humanitarian organizations in low-, middle-, and high-income countries across Australia, Africa, Asia, Middle East, Europe, and the Caribbean. Within the humanitarian sphere, she has worked predominately with the Red Cross and Médecins Sans Frontières (MSF), most recently as Medical Director for MSF (Amsterdam). Melissa currently works for the MSF Research Unit on Humanitarian Stakes and Practices based in Geneva.

Daniel Messelken is a research associate at the Center for Ethics at Zurich University and leader of the Zurich Center for Military Medical Ethics (www.cmme.uzh.ch). He also serves as Head Ethics Teacher for the Center of Reference for Education on IHL and Ethics of the International Committee of Military Medicine and is member of the Board of Directors of the International Society for Military Ethics in Europe (EuroISME). Dr. Messelken studied Philosophy and Political Science in Leipzig and Paris (1998–2004) and received his PhD in philosophy from the University of Leipzig in 2010. Besides military medical ethics, his main research fields include just war theory, the morality of violence, military ethics, and applied ethics, more generally.

Philisiwe Precious Ncayiyana holds a BPharm, UKZN (2001); PDBA, GIBS (2006); MBA-Finance, City University London (2008); and MSc Med-Bioethics and Health Law, Wits (2017). She is a chief pharmacist at 1 Military Hospital. Previously, he was deployed at the Mobile Military Health Formation as a staff officer for pharmaceutical support to external operations. She has over 20 years of experience in Pharmacy. She is a deputy chairperson for the South African Military Health Services (SAMHS) Antimicrobial Stewardship (AMS) Committee and SAMHS Formulary Committee. She is a former consumer representative at the United Nations Development Program's Standard and Labelling Project Steering Committee. Her interests include AMS, polypharmacy, bioethics and health law, and military medical ethics. She has presented in countries like Belgium, Botswana, Namibia, Switzerland and Zimbabwe. She also has television and radio experience as a guest and co-producer at the South African Broadcasting Corporation (SABC).

Leonard Rubenstein is Professor of the Practice in the Department of Epidemiology at the Johns Hopkins Bloomberg School of Public Health, and Interim Director of its Center for Public Health and Human Rights. He is also a core faculty member of the Berman Institute of Bioethics and the Center for Humanitarian Health at Johns Hopkins University. Prior to coming to Johns Hopkins, Professor Rubenstein was a senior fellow at the United States Institute of Peace and before that President of Physicians for Human Rights. He has been the recipient of numerous awards, including the American Public Health Association's Sidel-Levy Award for Peace. He is a graduate of Wesleyan University and Harvard Law School. He founded and chairs the Safeguarding Health in Conflict Coalition and is the author of *Perilous Medicine: The Struggle to Protect Health Care from the Violence of War* (Columbia University Press, 2021).

Wunna Tun, MBBS, Msc, MD serves in the frontline area of Myanmar. He was Secretary of JDN, World Medical Association. Dr. Tun led on drafting statement of WMA Resolution in Support of Medical Personnel and Citizens of Myanmar and contributed on drafting and revision of medical ethics in times or conflict and war, Physician Pledge, International Code of Medical Ethics, various WMA medical ethics declaration and resolutions. He chaired and moderated WMA Thematic Session on Warfare Ethics last April 2022 and published a report about it on *World Medical Journal*. Dr. Tun served on various leadership positions locally and internationally. Since the Military coup in Myanmar, he is living and caring for Myanmar people in an area where booms resound from nearby combat, fighter planes soar overhead where Myanmar military fires volleys of missiles from the sky to kill communities and innocent people.

Eva van Baarle is assistant professor military ethics and philosophy at the Netherlands Defense Academy. She teaches military ethics at the Netherlands Defense College: both staff courses and international courses. She also is a project leader and trainer in the train-the-trainer course military ethics for non-commissioned officers (all services). Her research focusses on fostering moral competence of military personnel through ethics education; reflection on hazing rituals in the Armed Forces; understanding moral issues in soldier enhancement; and she is involved in an action research project aimed at fostering a "just and psychologically safe culture in the Armed Forces."

Adriaan van Es received his training as a doctor at Medical School Utrecht from 1966–1973. Later, he was Medical Officer in Charge at Dormaa Hospital, Ghana (1975–1979), worked as a Family doctor in Amersfoort (1979–2015), and served as a GP-trainer at Medical School Utrecht (1990–2015). He also worked as a MD for asylum seekers (2015–2019) and as a MD at the Moria Refugee Camp, Lesbos, Greece in 2019. He was Co-founder and president of the Johannes Wier Foundation for health and human rights (1985–1998) and he is the Founder and secretray of IFHHRO | Medical Human Rights Network www.ifhhro.org (2002–).

He was also part of various fact-finding missions, reports, guidelines and articles as a member of the International Forensic Expert Group IFEG. He is a co-author of the Istanbul Protocol 1999 and 2022 as well as of the Dual Loyalty & Human Rights in Health Professionals Practice 2002.

Yuxuan Yang received her PhD in Public Health from the Academy of Military Medical Sciences. Her academic expertise includes military medical ethics, healthcare management, and health economics. She has worked on a military medical ethics project for three years and published a number of research articles.

Min Yu is a professor of Health Services Administration and Management in the Academy of Military Medical Science. He was appointed as associate professor in 1995 at the Fourth Military Medical University. He visited Harvard School of Public Health, Department of Health Policy and Management for one year in 1997. He was appointed as professor in 2000. He worked as Chief Force Medical Officer in the United Nation Mission in Sudan for one year in 2006. Since 2010, he has worked as the military medical expert of United States' Technical Advisory Group for the Pilot Project on Military Medical Support Capability Development. Since 2013, he has been working as a professor at the Academy of Military Sciences. From 2019 to 2022, he was the chairman of the Technical Commission on Administration and Medico-military Logistics of the International Committee of Military Medicine.

Chapter 1
Preface—In Search for the Meaning of Medical Neutrality

Ana Elisa Barbar and Daniel Messelken

Abstract This introductory chapter examines some of the interpretations of the concept of medical neutrality that it has received from a variety of authors. It distinguishes three frequently used approaches (or interpretations) of "medical neutrality". The first is neutrality as part of humanitarian principles to be applied in humanitarian action in general, while the second is linked to the exclusivity of the medical mission, ensuring the protected status of medical personnel and objects as non-combatants. The last approach is related to medical ethics and respect for human rights as obligations that medical personnel should fulfill. The introductory chapter aims to provide a framework for the subsequent discussion by outlining the key distinctions between the various contributions to the book.

1 Introduction

This book explores the various ethical implications of "medical neutrality" in the professional lives of medical personnel working in military or humanitarian settings, as well as some ethical challenges surrounding the concept of medical neutrality. As a starting point for the discussions proposed by those contributing to this volume, it is important to look at the various meanings that the concept of medical neutrality has received when interpreted or applied by different authors. Contrary to what one might think, medical neutrality is far from a unanimous concept, and the understanding of what it implies has drastic consequences for the practice of medical personnel.

A. E. Barbar
Insecurity Insight, Geneva, Switzerland

D. Messelken (✉)
Center for Ethics, ZH Center for Military Medical Ethics,
University of Zurich, Zürich, Switzerland
e-mail: messelken@ethik.uzh.ch

D. Messelken, A. E. Barbar (eds.), *Challenging Medical Neutrality*, Military and Humanitarian Health Ethics, https://doi.org/10.1007/978-3-031-69398-4_1

Uses of the concept of medical neutrality range from "not engaging in hostilities" and "exclusively assigned to medical activities" to meanings that conflate the concept with "impartiality" or even "adherence to human rights". Thus, the nuances of what is presented in guiding ethical frameworks, debated in technical texts, and expected of the behavior of medical personnel with respect to such neutrality can sometimes be quite confusing. A clearer understanding is highly needed in view of the frequent use of the term "medical neutrality" as an element of the protection of healthcare, but without a clear reference to what are (or could be) the actions coming from healthcare personnel to demonstrate or perform such neutrality. Moreover, recent scenarios of conflict, civil insurgence and instrumentalization of healthcare for other purposes call for an open debate on whether medical neutrality should be applied and when certain actions carried out by medical personnel might be perceived or interpreted as non-neutral, thus creating further dilemmas for the health personnel. To add to the confusion, "neutrality" might also be referred to as "medical sanctity" or "medical immunity", depending on the meaning attached to it—suggesting inviolable values that go beyond the instrumental approach of neutrality, often presented by humanitarians.

In the remainder of this chapter, we will present some of the different meanings attributed to the concept of "neutrality" in healthcare, as well as social and historical aspects associated with these different definitions.

2 Neutrality—From a Humanitarian Principle to Its "Medical Use"

The concept of "neutrality" has its origins in the idea of military neutrality, which means not giving an advantage to or supporting any warrying party, but standing outside of the conflict. It was present in the first Geneva Conventions of 1864 and could be attached to notions of permanent or temporary neutrality, depending on the context and on the object/person attached to the responsibility of neutrality.

Politically, "neutrality" has to do with a strategic choice to remain available or associated to different sides or interested parties, with an acknowledged risk of losing characteristic values.

In the humanitarian sector, influenced by the tradition of the principles initially set by International Red Cross and Red Crescent Movement, the most used definition comes from the Office for Coordination of Humanitarian Affairs (OCHA) of the United Nations, which identifies "neutrality" as one of the four core humanitarian principles, with the following definition: "*Humanitarian actors must not take sides in hostilities or engage in controversies of a political, racial, religious or ideological nature*". Of course, this principle speaks to all kinds of humanitarian action, and not to medical action alone, although it is this definition that is mentioned most often in health-related documents.

If this definition of "neutrality," referred to as a "principle," were adopted by the health community as it is by humanitarians, it would imply an ethical responsibility on the part of medical personnel in understanding the concept of "medical neutrality." (Leaning 1999) However, for most medical personnel, the set of humanitarian principles is not the primary ethical guideline. A key international reference regarding medical ethics is the International Code of Medical Ethics, a document under the responsibility of the World Medical Association (WMA). In this text, there is no mention of medical neutrality, as much as other policy documents from this organization (such as the Havana Declaration) refer to the topic. It is important to note that, just like the WMA, the International Committee of Military Medicine does not have an operational definition of the concept of medical neutrality, and the mentions found in different working and policy texts will result in the same multiplicity of meanings that will be debated ahead.

3 "Medical Neutrality" as a Concept Deriving from International Humanitarian Law

As noted above, an important meaning given to "medical neutrality" relies on the understanding of neutrality in International Humanitarian Law (IHL), which associates the protected status of health care personnel to a prohibition of their participation in the hostilities. Just as civilians, medical personnel are *non-combatants*, they must remain "neutral" regarding the military conflict. As early as 1863, in the first version of the Geneva Convention the term "medical neutrality" was used to indicate that ambulances and military hospitals should not be associated to combat activities, alongside the wounded, "official medical personnel", voluntary medical personnel and other civilians engaged in the care of the wounded.

This notion of neutrality seems to address military medical personnel in particular, as they are inevitably associated with a party to the conflict, but do not take part in hostile activities. In this sense, this definition adds an element of exclusivity of medical functions, meaning medical neutrality is, beyond the non-participation in hostilities, a dedication to medical activities only, with a responsibility to provide care according to medical ethics. Nonetheless, the exclusivity-approach was only made explicit in later versions of the Geneva Conventions, when the specific mention of the term "neutrality" had already disappeared from the text.[1]

Still regarding the idea of neutrality as non-engagement in the hostilities and the prioritization of medical activities, it is important to acknowledge that it imposes a responsibility that goes beyond the control or direct action of medical personnel: it requires combatants to respect such neutrality and to prevent any action that might hinder the delivery of medical care, such as attacks against healthcare or blockages of access. Some attempts to define the term "medical neutrality", as seen in

[1] Interestingly, this notion of "medical neutrality", precisely because the term has disappeared from the text of the Conventions, is not a legal definition, and is based mainly on interpretations of IHL provided by the International Committee of the Red Cross.

open-source collaborative platforms or in the work of important organizations refer rather to this "third-party responsibility" and focus the meaning of "medical neutrality" on the respect and non-violation of the protected status rather than on the duties of medical personnel to uphold such neutrality (World Health Organization - WHO 2013; Mullan 2016). For example, in its fact sheet "Introduction to Medical Neutrality", the organization Physicians for Human Rights defines it as the "*principle of non-interference with medical services in times of armed conflict and civil unrest*" (Physicians for Human Rights - PHR 2013).

In the same line of argument, terms such as "sanctity of medical/ health care", "inviolability of the medical mission" or "protection of the medical mission" have been associated with this interpretation of neutrality of medical care as protection from violence, but beyond contexts where IHL is applicable. While borrowing some of the meaning brought in by IHL, they speak of keeping medical care as "untouched" by political, moral or social disruptions, or of preventing any violence that would disturb health care, with the goal of protecting the delivery of care and maintaining life-saving efforts even amidst insecure, fragile or hostile environments (Nelson Mandela Foundation 2017). Also in the legal area, the notion of "neutrality" is applied to denote potential hindering effects of sanctions, or to argue against the multiplication of military or defense-related actions that associate health care with military or counter-terrorism actions (Bouchet-Saulnier 2021).

4 "Not taking sides": A Disputed Approach to Neutrality, Beyond Engagement in Hostilities

Another approach to "medical neutrality" associates the ethos of a careful and balanced decision-maker with the figure/ role of the health care practitioner. This interpretation suggests the achievement of a neutral position as an ability to overcome any individual traits such as race, heritage or ethnicity, or potential biases linked to political or religious beliefs. Beyond the aforementioned prohibition to engage in disputes or hostilities, and the exclusive focus on medical activities, the understanding of medical neutrality as an absolute dissociation of the health care personnel from any elements of public dissent seems unrealistic for various authors (Rubenstein 2013).

In some lines, such an interpretation of the concept of "medical neutrality" hints to the responsibility of health care personnel to respect medical ethics, and thus to be "neutral" to other influences that could disturb or hinder the ethically ideal provision of care. This could also appear as a "value-neutral" approach to care, based solely on medical ethics and scientific evidence, without acknowledging that both medical ethics and scientific statements can hardly be separated from disputes that surround structural power dynamics in society (Gasparetto et al. 2018).

The reasons for disagreement with this interpretation of an absolutely "anti-partisan" or value-neutral approach, as well as the debate around its limits and risks, are the subject of several chapters in this book and will be explored through different angles.

5 "Medical Neutrality" as Respecting Human Rights

In recent debates, some authors present "medical neutrality" as adherence to the values put forward by human rights (Physicians for Human Rights PHR 2017; British Medical Association (BMA) 2022). In this approach, a "neutral" health professional would strive to respect human rights, with a strong semantic proximity to the idea of maintaining impartiality in care.

While under the same umbrella of arguments, the concept of "medical neutrality" is sometimes contested in relation to what "neutral behavior" could mean in the face of human rights violations. Those who worry that "neutrality" would be tantamount to negligence or inaction point out that, in addition to respecting medical ethics, health personnel should also respect the rights set forth in the International Human Rights Framework and other international instruments, such as the Convention against Torture and Other Cruel, Inhuman or Degrading Treatment or Punishment. This view implies that health workers have a responsibility to "take sides", i.e. to avoid neutrality, and to protect the dignity of those against whom violations are being committed—even when the victims of such violations are not directly their patients (World Medical Association 2022).

This understanding of "medical neutrality" as a responsibility to take sides when faced with human rights' violations clearly diverges from the one attached to the humanitarian principle, since it requires (rather than prohibits) the involvement of health personnel in certain disputes that may be considered political or ideological, as far as they should denounce and prevent any violations of human rights. Beyond their exclusively health-related responsibilities, this "medical neutrality" as respect of human rights also implies that healthcare workers must pro-actively avoid any discrimination[2] and they should report whenever they witness a problem in this regard. This interpretation might suggest a broader societal role of health care personnel, who are inevitably part of a power dynamic determined by historical, social, economic and hierarchical structures, especially in the case of military health workers or health workers associated with a national government program.

6 Overlaps Between "Neutrality" and Other Ethical Principles in Healthcare

Another challenge to understanding the concept of "medical neutrality" consists in the overlap of medical neutrality with a different set of ethical prerogatives. From WHO documents to reports by various organizations,[3] "neutrality" and "impartiality" are used with almost interchangeable meanings (Nathanson 2013). This seems

[2] Here as in discrimination regarding one's human rights, not as in medical discrimination for care.

[3] For an example, see the interchangeable use in material promoted by the organization Physicians for Human Rights: https://phr.org/issues/health-under-attack/medical-impartiality/

to be detrimental to a proper differentiation of the two principles, but also to understanding their singular importance and requirements to behave in ways that conform to the different goals of each principle. Impartiality, a concept much less disputed than neutrality, refers to the approach to clinical care: the provision of health care should, at all times, respect the primacy of health (clinical) needs, and avoid any other parameter of prioritization in decision-making.[4] Thus while "impartiality" refers to the way in which care should be provided, "neutrality" speaks of protecting or prioritizing the provision of care in the face of other contextual elements.

In addition to being conflated with "impartiality," the term "medical neutrality" is also used to indicate the application of the ethical principles of beneficence, to suggest autonomous or independent decision making by the health care professional, and to refer to the equitable distribution of health care resources (Peppin 1995; Kloiber 2013; Sulmasy et al. 2019).

7 Conclusion

We have distinguished three frequently used approaches (or interpretations) of "medical neutrality". The first one is neutrality as part of humanitarian principles to be applied in humanitarian responses at large, while the second one is tied to the exclusivity of the medical mission, assuring the protected status of health care personnel and objects as non-combatants. The last approach is related to medical ethics and respect for human rights as obligations that medical personnel should fulfill.

Despite the challenges of finding a common definition, the following chapters will approach different angles of a complex discussion, mixing the practical experiences of frontline military and civilian health care workers with the academic work of various contributors.

8 Synopsis

To conclude this introduction, we will present the chapters of the book in order to highlight the main lines of argument of the volume. The contributions to this edited volume come from a variety of backgrounds and include historical, practical, and more theoretical analyses of the overall topic of medical neutrality.

In the **first part of the book**, two historians offer insights into precursors of the current debate and show how the discussion on the meaning of medical neutrality goes back a long way. **Duncan McLean** sheds light at the history of attacks on health care during conflict, which is the flip side of not accepting their neutrality.

[4] Some authors also call this principle "non-adverse discrimination", that is, the only discrimination possible is that based on clinical needs, and in benefit of the person who will receive care.

The paper puts these current events in historical perspective with the aim of better understanding contemporary attacks on health care. The issue of non-respect of the protection of healthcare is seen as one of the big challenges in contemporary warfare and has been analyzed by several organizations in recent years. However, a historical comparison is missing from the debate and this chapter fills that gap. The author draws on his experience both as an analyst with *Médecins Sans Frontières* (MSF) and a historian to provide concrete illustrations.

In the second contribution on historical aspects, **Adriaan van Es** looks at dual loyalty problems of health professionals, which arise when medicine cannot be practiced independently of the circumstances and thus not in a neutral way. He analyzes the issue in the context of a human rights-based approach and uses historical examples of medical involvement and complicity in violations of human rights, as well as the mobilization of medical knowledge in the defense and promotion of human rights. The aim of his chapter is to provide guidance, based on lessons learned from the past, to assist health care personnel in meeting future challenges. Van Es stipulates that health professionals have a duty to denounce human rights violations which can be grounded in the neutrality of their profession, which makes the obligation to their patients the most important one.

The **second part of the book** is devoted to some more theoretical reflections that lay the groundwork for the more practical contributions that follow in the next section. First, **Eva van Baarle** examines how the moral value of humanity underlies the principle of medical neutrality and elaborates on the consequences of whether it is considered to be an absolute moral value. If it is an absolute moral value, then it might still have to be weighed against practical restraints in military or humanitarian missions, leaving health care personnel in such situations with immensely difficult choices, as they must evaluate both moral questions of guilt and medical questions of treatment priorities.

Erik Gustavsson looks at the question of prioritizing and rationing and asks whether medical neutrality could be a ground for rationing medical care in the context of armed conflict. Hence, medical personnel should not discriminate on "morally irrelevant grounds" such as age, gender, nationality, and political affiliation. While such a principle specifies what cannot be a reason for distributing resources, it does not provide positive guidance on what does. Gustavsson argues that medical neutrality as impartiality must be understood as a mid-level principle, rather than a high-level principle, to be applicable in practice. He discusses how the application of such principles might work in practice and discusses some of the issues that might arise.

Paul Gilbert discusses moral arguments for the principle of medical neutrality. In response to the suggestion that the military role of military physicians might sometimes lead them not to act neutrally, Gilbert argues that such a position misunderstands the role of physicians in the armed forces. He discusses an alternative approach in which the armed forces would not employ physicians bound by medical ethics, but rather "medical technicians" who could follow military orders without being bound by medical ethics. Gilbert concludes, however, that such personnel would normally have to follow the same priorities as full medical personnel in adhering to the principles of IHL. Thus, the paper ends with an argument about the importance of maintaining the current idea of medical neutrality.

The **third part of the book**, which is the most comprehensive one, collects case studies and more applied analyses of how medical neutrality is implemented and criticized or ignored in reality.

Martin C. M. Bricknell, who served in the British Armed Forces for over 30 years, discusses the tension between health services operating under state authority (and therefore primarily responsible to their citizens) and those operating under humanitarian principles, such as UN agencies and NGOs. He argues that government health services can neither be neutral nor operationally independent, but remain bound by the principles of humanity, impartiality, and international humanitarian law. He suggests replacing the term medical neutrality with medical humanity, as this better captures the relevant aspects and does not presuppose a neutral position of health care personnel.

In their chapter, **Min Yu** and **Yuxuan Yang** analyze the unique role of military health care personnel in UN Peacekeeping Operations, sharing insights from a long career in military healthcare. While UN peacekeepers often come from diverse backgrounds, as peacekeepers they follow three basic principles: consent of the parties, impartiality, and non-use of force. The meaning of impartiality has not always been clear, but it is now understood to require impartiality in dealing with the parties to the conflict, but no longer neutrality in the execution of their mandate. The authors discuss how this understanding affects the medical component of peacekeeping missions and illustrate their findings with case examples.

Focusing on garrison health care, **Innocent Okafor Eze** and **Uchenna Clara Innoeze** reflect on the conflict between medical ethics and the military demands on the physician-soldier. The ethical challenges in the daily work of military doctors in occupational healthcare have been neglected and are still underdeveloped compared to debates on ethical challenges in armed conflict. Yet, the concepts of loyalty to superiors and organization in the military and loyalty to principles and patients in the medical profession are very different. Drawing on their own experience as a military doctors, the authors identify conflicts that threaten the ethical demands of patient autonomy, abuse of justice and neutrality, scientific validity, freedom of conscience and professionalism. They conclude by suggesting ways to better resolve these tensions and emphasize the importance of the independence of the medical profession within military organizations.

The case of Dr. Wouter Bason in South Africa was a highly publicized case of the use of medical knowledge in military operations for non-medical purposes. **Philisiwe Precious Ncayiyana** presents the case and its different angles explored by the institutions responsible for judging the case at the time. Using both documentary sources and interviews with members of the South African Special Forces, the author also proposes her own analysis of the implications of Dr. Bason's case for medical ethics. She suggests that there may be a need to develop specific ethical guidelines for military doctors and to include military medical personnel in civilian-led trials for ethical misconduct of military health personnel.

Recounting his own experiences after the military coup in Myanmar in 2021, **Wunna Tun** discusses what medical neutrality means from the perspective of a humanitarian health care provider in such difficult circumstances. On the one hand, he narrates the numerous attacks against health care providers and how the medical profession has come under attack. On the other hand, he also critically reflects on his own experiences in an actual combat zone and on how medical neutrality can be implemented.

While economic sanctions are often seen as a good tool for non-violent action, their humanitarian impact and their potential to damage health systems is also often overlooked. **Julian W. März** analyzes the implications of the human right to health and the principle of medical neutrality for the use of economic sanctions and explores how unintended side effects of economic sanctions may affect both the right to health and medical neutrality even in third countries. The author also presents possible solutions to mitigate the unintended humanitarian impact of economic sanctions.

The **fourth part of the book** takes a broader perspective, opening the debate on how medical neutrality can (or cannot) respond to the needs of medical personnel and the challenges of contemporary health care settings.

In her chapter **Melissa McRae** explores the impact of leveraging health for other outcomes, using as a case the Global Health and Peace Initiative, currently under debate at the World Health Organization. She suggests that efforts to achieve non-medical goals, exploring the medical mission and its ethical limits, have had a detrimental effect on the acceptance of medical personnel and medical interventions, particularly in humanitarian settings. She concludes by pointing to specific risks associated with such efforts, including the loss of medical neutrality in contexts where such leveraging of health interventions might be perceived by parties of a conflict as political positioning of the medical personnel.

Concerned by the lack of clarity about what is expected of health personnel when upholding neutrality, **Ana Elisa Barbar** discusses the potential risks associated with the semantic diversity of the term. Using various scenarios, from peacetime to conflict settings, she explores the consequences of an undefined concept of medical neutrality on the individual well-being and up to professional decision-making and policy-making layers. In her conclusion, she argues that, beyond set obligations to the parties of a conflict, medical personnel should have clear instructions and expectations so that upholding such neutrality is a behavior more clearly understood among stakeholders.

Still within the perspective of multiple meanings ascribed to medical neutrality, **Len Rubenstein** advances the idea that medical neutrality may simply be unattainable if understood as a prohibition agains engaging in all polemics of a political nature. In this closing chapter, the author presents a historical approach to understanding how medical neutrality cannot be reduced to such an admonition, arguing that medical personnel should in fact be encouraged to defend certain values that support humanity and protect the rights of the patients and of larger populations, and that ignoring or failing to acknowledge certain aspects of the context in which care is provided might violate core ethical principles in medical ethics.

Acknowledgements The authors acknowledge the support of the Centre of Competence for Military and Disaster Medicine, Swiss Armed Forces, Bern, Switzerland.

References

Bouchet-Saulnier, Françoise. 2021. How counterterrorism throws back wartime medical assistance and care to pre-Solferino times. *International Review of the Red Cross* 103: 479–516. https://doi.org/10.1017/S1816383121000771.

British Medical Association (BMA). 2022. *Health and human rights in the new world (dis)order*. London: BMA.

Gasparetto, Alessandra, Ralf J. Jox, and Mario Picozzi. 2018. The notion of neutrality in clinical ethics consultation. Philosophy, Ethics, and Humanities in Medicine 13: 3. https://doi.org/https://doi.org/10.1186/s13010-018-0056-1.

Kloiber, Otmar. 2013. *Professional autonomy and self-governance*. Remarks by the WMA Secretary General. WMA.

Leaning, J. 1999. Medicine and international humanitarian law. *BMJ* 319: 393–394. https://doi.org/10.1136/bmj.319.7207.393.

Mullan, Zoë. 2016. Medical neutrality: Resetting the moral compass. *The Lancet Global Health* 4: e215. https://doi.org/10.1016/S2214-109X(16)00069-3.

Nathanson, Vivienne. 2013. Medical ethics in peacetime and wartime: The case for a better understanding. International Review of the Red Cross 95: 189–213. https://doi.org/https://doi.org/10.1017/S1816383113000775.

Nelson Mandela Foundation. 2017. *Guidelines for the provision of healthcare services during higher education protests*. Johannesburg: Nelson Mandela Foundation.

Peppin, John F. 1995. Physicians' values and physician-value neutrality. *Journal of Religion and Health* 34: 287–300. https://doi.org/10.1007/BF02248738.

Physicians for Human Rights PHR. 2013. *Introduction to medical neutrality. Fact Sheet*. PHR.

———. 2017. *Medical neutrality and the right to health. Policy Brief*. New York: PHR.

Rubenstein, Leonard S. 2013. A way forward in protecting health services in conflict: Moving beyond the humanitarian paradigm. *International Review of the Red Cross* 95: 331–340.

Sulmasy, Daniel P., Ilora Finlay, Faith Fitzgerald, Kathleen Foley, Richard Payne, and Mark Siegler. 2019. Physician-assisted suicide: Against medical neutrality. *Journal of General Internal Medicine* 34: 1372–1372. https://doi.org/10.1007/s11606-019-05019-1.

World Health Organization (WHO). 2013. *WHO condemns attack on health care facility in Las Canood, Somaliland*. WHO Press Release.

World Medical Association. 2022. *WMA international code of medical ethics*. Geneva: WMA.

Part I
Medical Neutrality in Historical Perspective

Chapter 2
Historical Antecedents to Understanding Contemporary Attacks on Healthcare

Duncan McLean

Abstract The protection of medical services in armed conflict is generally framed under the principle of medical neutrality. That respect for this principle has had a chequered past would be an understatement, that violations of medical neutrality continue to this day obvious to most observers. Rather than attempting to explain why this is the case, this paper offers a comparative analysis as to how a historian's toolbox can be employed to better understand contemporary attacks on healthcare. A broad historical overview of medical neutrality will be provided, while remaining cognizant of the challenges of juxtaposing different epochs, before delving into methodological suggestions. Throughout the paper the author draws on his experience both as an analyst with Médecins Sans Frontières (MSF) and a historian to provide concrete illustrations, both historical and recent. Finally gaps and suggestions for future research will be highlighted.

In an operational capacity for *Médecins Sans Frontières* (MSF), the present author had the opportunity to visit the organization's medical activities in the Nuba Mountains in early 2013. Geographically situated in Sudan's South Kordofan province, armed resistance from the Sudan People's Liberation Movement's northern brethren had continued despite South Sudanese independence in 2011. The reaction from Khartoum included the bombing of combatants and civilians alike.

While aerial bombardment has been portrayed as sanitized warfare, the view from the ground is obviously far different. In the MSF-supported hospital in Farandalla

The premise of this chapter was presented at the conference *Rethinking the History of Violence, Health and Care in Wartime, c. 1860–2000s*, organized by the University of Manchester's Humanitarian and Conflict Response Institute. The views expressed are those of the author and in no way represent the organization to which he belongs.

D. McLean (✉)
Unité de Recherche sur les Enjeux et Pratiques Humanitaires (UREPH), Médecins Sans Frontières – Switzerland, Geneva, Switzerland
e-mail: duncan.mclean@geneva.msf.org

D. Messelken, A. E. Barbar (eds.), *Challenging Medical Neutrality*, Military and Humanitarian Health Ethics, https://doi.org/10.1007/978-3-031-69398-4_2

village, basic precautions had been taken. Trenches and foxholes were dug in the immediate vicinity. When the highly recognizable drone of a Soviet-made Antonov was heard, time permitting, patients and staff were evacuated underground. Barrel bombs were the preferred ordnance, essentially highly imprecise "improvised containers" filled with explosive material (Weapons Law Encyclopaedia 2013). Local wisdom contended that once sheltered you would be extremely unlucky to experience a direct hit, burst eardrums being the more common eventuality.

Luck for Farandalla's sole medical facility ran out the following year when two bombs hit the hospital on 16 June 2014, wounding several patients and an MSF staff member (MSF 17 June 2014). A little over 7 months later, the same structure was "directly targeted" by the Sudanese Air Force with a cluster of 13 bombs landing inside and outside the hospital compound, again resulting in injuries but no fatalities. On both occasions the authorities in Khartoum were aware of "MSF presence and activities in the hospital." And while MSF publicly condemned the attacks there were also practical consequences with the second incident leading to the organisation suspending medical activities in South Kordofan, depriving the population of one of the few healthcare providers in the region (MSF 22 January 2015a).

The indiscriminate bombing campaign in the Nuba Mountains, and the notable targeting of Farandalla's hospital, is hardly exceptional as concerns the protection of medical services in armed conflict. From the admittedly limited prism of MSF's experience over the past decade, colleagues, patients, caretakers, and associated staff have been killed in Ethiopia, Afghanistan, South Sudan, Syria, Yemen, Central African Republic, Burkina Faso and, most recently, the Occupied Palestinian Territory. In all cases, attempts are made to "establish the facts on what happened and why, and to establish whether we can put teams back on the ground" (MSF 8 June 2021b).

But there is also a bigger picture. That respect for the principle of medical neutrality has had a chequered past would be an understatement, that violations of medical neutrality continue to this day obvious to most observers. But while identifying explanatory factors can be elusive, attempts at understanding the nature of each attack relevant to each context can be revealing. This chapter offers a comparative analysis as to how a historian's toolbox can be employed to better understand contemporary attacks on healthcare. A broad historical overview of medical neutrality will be provided, while remaining cognizant of the challenges of comparing different epochs, before delving into methodological considerations. Concrete illustrations, both historical and recent, will be provided throughout the narrative before highlighting gaps and suggestions for future research.

1 Lieber's Long Shadow

It might seem common sense that the hospital in Farandalla was operating under a banner of medical neutrality, but this should not be taken for granted. Protected status can be revoked if "used to commit, outside their humanitarian duties, acts harmful to the enemy", a fundamental tension when balancing military and humanitarian prerogatives. But even then, exceptional circumstances aside, such protection can only cease "after due warning has been given", including the provision of a "reasonable time limit." This was not the case in the described example, a clearly designated civilian hospital "organized to give care to the wounded and sick, the infirm and maternity cases" (Convention IV, articles 18 and 19). However, in a confrontation between international humanitarian law (IHL) and the cynical reality of what can be described as military necessity, the latter took precedence irrespective of the existing statutes.

Leos Müller (2019) has described neutrality as a "strange thing", a concept that has frequently been dismissed as "either unrealistic or immoral – or both." It certainly has a long history with origins that predate the modern era. As with other aspects of international law such as diplomatic immunity and formal treaties, its roots can be traced "back to the days of Rome and Greece, or even earlier times" (Jessup and Déak 1935, 3–4). As a foreign policy option, neutrality has been particularly attractive to smaller states, a guarantor of sorts for their sovereignty and independence. It remains a "key concept of international law" today, playing a central role in the rise of "internationalism, humanitarianism and the peace movement, and so has contributed to the foundations of twentieth-century international organizations such as the League of Nations and the United Nations" (Müller 2019, 3).

While state neutrality essentially means not taking sides in a conflict; avoiding military alliances; and not taking actions that would advantage one side or another, applying this concept to aid organizations requires a different approach. For humanitarian neutrality to function all sides must accept that relief actions are "not hostile acts, nor are they de facto contributions to the war efforts of one of the belligerents." This includes humanitarian personnel who are expected to "abstain from any direct participation in hostilities or committing an act harmful to a party to the conflict" (Bouchet-Saulnier 2015). The application of medical neutrality follows a similar logic: not only non-interference in medical services during armed conflict, but those same medical services must ensure the impartial treatment of the sick and wounded irrespective of their provenance.

Tracing the origins of medical neutrality in the codified legal sense invariably involves returning to the mountain of literature produced since the First Geneva Convention of 1864. De Baets has argued that prior to this foundational moment, the history of humanitarianism was "often neglected apart from some occasional philosophical musings about the bellicose nature of human beings or the ritual nod to the 1648 Peace of Westphalia." But if IHL is indeed divided into "before and after Henry Dunant", a contemporary of the driving force behind the establishment of the Red Cross merits similar attention albeit for very different reasons (2022, 1588). Francis Lieber explored and then advocated for the application of military necessity when articulating his vision of the rules of war, including the protection of medical services.

In the context of IHL, the concept of military necessity posits that rapidly achieving battlefield objectives should be a priority even when this at times necessitates extreme levels of violence. Commissioned to draft *Instructions for the Government of the Armies of the United States in the Field*, Lieber completed his work the year before the First Geneva Convention was enacted. The parallels and divergences are striking. In his detailed study, Leonard Rubenstein (2021, 23) notes that Lieber's code encouraged the protection of hospitals but only insofar as "the contingencies and the necessities of the fight will permit", the "necessities of the fight" essentially permitting both the denial of care and the destruction of medical facilities. This dichotomy of visions has been summarized as Dunant's "absolutist protective stance" versus Lieber's view that compassion for the casualties of war must be "tempered by the need to win a just war, and quickly – that is, military necessity" (Rubenstein 2021, 34).

However much rejected in law, Lieber's arguments have maintained an outsized influence in military doctrine, understandably terrifying to the humanitarian sector. The most recent incarnation of the Geneva Conventions remains the only international legal framework that aid organizations can draw on to "justify their own presence in war zones and make demands for access and protections" (Hofman 2021, 204). This precarious legitimacy is placed fundamentally at risk by actions that closely resemble a nineteenth century interpretation of military necessity, a concept that "easily morphs into military convenience" (Rubenstein 2021, 58). It is difficult to understand and interpret at least some of the attacks on healthcare today without integrating the logic of Lieber's long shadow.

2 Mitigating Historical Pitfalls

Even with a clear understanding of medical neutrality and its applications, infractions are difficult to compare in historical terms. The Geneva Conventions themselves, along with contemporary interpretations under customary law, can be described as living documents, their many incarnations reflecting both the changing means of waging war and attempts to broaden the scope of safeguards available. The eventual incorporation of non-international armed conflicts into the IHL protective framework being an obvious case in point. There are however other pitfalls to comparative analyses when investigating violations of medical neutrality that merit review.

Juxtaposing historical epochs, whether in terms of vastly different contexts or attack typologies, raises questions of comparison over space and time which are extremely difficult to answer, especially in quantitative terms. This includes data collection which can vary within organizations let alone across the humanitarian sector, even today. Such challenges are compounded by the increase in the number of aid agencies which can have a direct impact on levels of exposure and the frequency of incidents. Determining whether an attack was deliberate, dismissive of basic precautions, or unfortunate collateral damage is similarly challenging and often impossible. Finally, different forms of public pressure, notably through the rapid availability of information and disinformation, can skew analyses.

Given the above constraints, clarifying the investigation parameters is paramount. To return to the prism of MSF's experience, in the event of an attack on medical facilities, personnel or patients, some form of denunciation will almost certainly appear, often followed by "calls for an independent investigation." However, an internal analysis is likewise essential to both resist and, if necessary, challenge the "narrative and facts" put forward by various actors in a conflict in addition to the media and other interested parties. Another important objective is to improve the security measures of existing operations, including the identification of potential blind spots as per MSF's protected status, and to feed further negotiations to secure stronger agreements with the belligerent parties. In all cases, "exhaustiveness" is essential to avoid "subjectivity, presumptions and bias." Elements that can be asserted with a "high level of certainty should be presented as such", adding to the credibility of conclusions that could cause discomfort or embarrassment to those responsible for an attack, intentional or otherwise (MSF Minimum standards 2016).

With those objectives in mind, useful recourse to historical methods becomes self-evident, especially when a broad-brush search for long-term trends is replaced with a more granular and context-driven approach. Whether exposing a long-forgotten attack on a medical structure or a recent violation, basic archival investigative work is similar in attempting to reconstruct events as factually as possible. This includes the use of formal primary sources, diplomatic and legal documentation for example, and recourse to informal exchanges, such as personal letters and memoirs versus twenty-first century communication. Similarly, context and risk analyses need to be based on information available at the time, including the identification of potential patterns of attack, given the risk of retrospective bias, essentially unintentionally passing judgement through the subjectivity of hindsight, irrespective of the time interval.

There are a few obvious differences that can be emphasized. When comparing historical and recent attacks on healthcare, an inverse form of confidentiality is at play as concerns obtaining sensitive material. Access to the archives of the International Committee of the Red Cross (ICRC) provides an apt demonstration, highly accessible pre-1975, far more restricted afterwards. And while there is no possibility to interview long-deceased actors from historic attacks, there is a forensic reality in contemporary cases that can necessitate extreme measures to protect the identity of those involved, particularly victims and their families. Finally, there are limits to what can be gleaned from the historical record. Drawing lessons from a recent attack can have a level of pragmatic urgency, both to avoid a repetition and to adjust security policies accordingly.

3 Past Cases and Perpetual Dilemmas

It is this last point that becomes particularly relevant, and tricky to determine, in both historical and contemporary analyses of medical neutrality in practice. Essentially the process and degree to which codified ideals and aspirations perform in actual conflict zones. In this regard, Larry Minear (1999, 68) has suggested that the "real meaning of neutrality is tested where international humanitarian law and principle encounter the real-world dilemmas in the field." The historical record contains mixed results.

To illustrate these mixed results, the narrative now turns to earlier research by the author on the history of attacks on medical care. Recognizing the difficulties of comparing historic challenges of protection with contemporary incidents, similarities and differences were sought in the public explanations and rationales employed by those responsible for attacks on health facilities and staff. The objective being to see if elements of value relevant to continued attacks today would emerge. Framed as "Medical Care in Armed Conflict: Perpetrator Discourse in Historical Perspective", specific incidents over a wide range of contexts were reconstructed, primarily from first-hand sources, much as previously described.

The analytical framework of perpetrator discourse was based on more recent MSF experiences of attacks on healthcare, some of which will be detailed further on. Broadly these can be summarized as remaining silent; not taking a position, stalling, or avoiding the discussion; admitting involvement, generally with an apology or justification; or denial of involvement, often with some form of rejection narrative. This framework was applied to attacks on medical facilities and ambulances during the Franco-Prussian War from 1870–71; the sinking of hospital ships during World War I from 1914–18; the targeting of Red Cross field hospitals during the Second Italo-Ethiopian War from 1935–37; the occupation and destruction of medical facilities during the Second Sino-Japanese War from 1937–45; and finally multiple instances of public and Red Cross medical structures destroyed during the Nigerian Civil War from 1967–70 (McLean 2019).

In all five case studies a pattern of sorts emerged. In the early phases of a conflict, admissions of responsibility were common, often accompanied by claims of accidental or unintentional targeting of medical care. This was particularly obvious in the torpedoing of hospital ships in the First World War. The Germans relayed an apology for the near-miss of a British hospital ship via their Washington embassy in early 1915 (McGreal 2008, 28–32). Two years later the German narrative had shifted significantly, with regular claims that the British were "violating the Hague Convention regarding the application of the Geneva Convention to maritime warfare" (Diplomatic Correspondence 1918). The accusation of transporting munitions and troops was a convenient justification for the continued targeting of hospital ships in a period of unrestricted submarine warfare.

Allegations that the enemy was abusing the protections granted under the Geneva Conventions and thus jeopardizing the medical neutrality of their health facilities were likewise common to all the case studies. Whether factual or contrived, supposed infractions were used to justify further attacks on hospitals. This then became part of broader wartime propaganda by all sides, reinforcing a dehumanizing rhetoric familiar to most conflicts. In the case of Ethiopia, the Italians integrated broader colonial attitudes into their propaganda. Not only was the Ethiopian resistance abusing the protections granted by medical neutrality, but they were also "too backward to be able to respect emblems" of the Red Cross (Baudendistel 2006, 116). The ICRC was not immune in this regard, suggesting that abuses in the Sino-Japanese

War were to be expected on both sides as the "mentality of Orientals" meant they were "incapable of our way of thinking" (ICRC 1937).

The most persistent challenge for medical neutrality, and the most striking element to emerge from the archival material, was countering the logic of military necessity. Irrespective of the rationale presented, humanitarian principles were ultimately expendable when military objectives loomed larger. During the Biafran War, exasperated officials went so far as to question the relevance of "a Convention which is now somewhat out of date", the 1949 incarnation not even 20 years old at the time (Letters from Overseas Council 1968). And in every example studied from the Franco-Prussian War to Nigeria in the 1960s, there were instances of the Red Cross emblem being removed as a risk reduction strategy.

4 Utility and Limits

The point of the very brief summary of selected historical instances of hospital attacks is not to promulgate cynicism over the effectiveness or relevance of IHL today. Protection norms have always been contested and ensuring respect for medical neutrality is far more difficult in practice than in words. Rather, it is to present some of the research angles and related methodologies employed in making sense of past cases, and to broach their applicability to better understanding attacks that continue to this day. Returning to MSF's experience, as we have seen, internal reviews are conducted after major incidents to "gather together the available information, to establish our own version of the facts and to form an opinion" (Mendiharat 2017). A number of examples point to both the utility and limits of such exercises.

On 3 October 2015 a United States gunship fired 211 shells at an MSF-run trauma centre in Kunduz, north-eastern Afghanistan. Including staff, patients and caretakers, at least 42 people were killed, and a further 37 people were injured. Doubting the veracity of the US and Afghan-conducted inquiries, MSF called for an independent and impartial investigation by the International Humanitarian Fact-Finding Commission. The states involved must acquiesce for such initiative to be launched and in the case of Kunduz this was refused. MSF's own internal investigation, retracing the exact circumstances of the attack, was thus essential as there was a public narrative that shifted in the days and weeks afterwards, moving from collateral damage and self-defence to an accident. Beyond the specifics of Kunduz, invaluable lessons on deconfliction were likewise developed (MSF 2015a, b).

In 2015 and 2016 a number of MSF-supported structures in northern Yemen were hit by the Saudi-led coalition. The attack on Shiara hospital on 10 January 2016 resulted in 6 deaths and 8 injured; 8 months later Abs rural hospital was hit, killing 19 people including an MSF staff member, and injuring a further 24. As with other serious incidents an internal review took place, including interviews of key staff and eyewitnesses combined with the analysis of internal and external documentation along with photographic material.

Even with the burden of proof lying with the party launching an attack, determining the circumstances and protection status remain key. In the case of Shiara hospital, MSF could state that there was no evidence to "suggest that the hospital was being used for any military purposes that would warrant the loss of its protected status under international law" (MSF 15 March 2016a). As for Abs hospital, it could again be asserted that the "neutrality of the hospital had not been compromised, the hospital had not been hosting any political or military activities and a strict no weapon policy was being applied" (MSF 26 September 2016b).

A final example concerns the 12 May 2020 massacre at Dasht-e-Barchi hospital, a maternity primarily catering to the minority Hazara community in Kabul. 24 people, mostly expectant mothers along with an MSF midwife, were "deliberately and methodically" killed. The internal investigation again drew on witness statements and analysis of key documentation to develop a detailed description and chronology of the attack. Important information was gleaned from the efficacy of existing security protocols (90 staff were able to take refuge in saferooms) but, despite speculation in the media, there was no confirmation of the culprit. With no claim of responsibility and no public narrative to potentially counter, there were clear limits on what could be learned, including how the medical facility's neutral and protected status was compromised to such an extreme irrespective of the rationale employed (MSF 10 May 2021a).

If the methods of analysis for past and present attacks are similar, it is worthwhile returning briefly to an important and already noted difference, the level of urgency. Establishing the "facts and identifying those responsible" for an attack has a direct impact on the continuation of care, demands for justice, and eventually compensation. Identification of the perpetrator is key in this regard, a task further complicated when perpetrators "deny, refuse or downplay their responsibility" (Mendiharat 2017). The historical record might indicate patterns of behaviour and the perennial shadow of military necessity but leaves little room for optimism in terms of accountability.

5 Research Gaps

There was certainly no accountability in the bombings of the Farandalla hospital. Suspicion fell on the Sudanese Air Force, quite simply because it was the only armed actor in that particular theatre of operations with that particular capacity. But if holding perpetrators to account can seem hopeless, it does point to an area of research that is relatively unexplored.

The Leipzig war crime trials of 1921 were an exception in this regard, a rare attempt to hold those responsible for an attack on medical care accountable. A German submarine commander was charged with ordering the sinking of the HMHS *Dover Castle*, a hospital ship, and the subsequent machinegunning of survivors. The defence pointed to military necessity, killing the "men and women in the lifeboats in order to prevent them from reaching their homes and re-joining the war against the Fatherland" (McGreal 2008, 225–26). Reviewing past and present cases while integrating perspectives of military necessity and the impact on medical neutrality could still have value.

Other areas to consider could include normative and especially pragmatic attempts to preserve neutral medical space, negotiation strategies employed, and possible mitigation measures. The direct and indirect impact of attacks on medical structures, including broader questions over subsequent access to health and corresponding links to mortality and morbidity, likewise merits attention. And finally, reflections that focus exclusively on violations of IHL, as with this paper, have an inherent bias by focusing exclusively on tragic events that invariably draw the eye. Research highlighting effective strategies at maintaining a protected space for medical care in times of conflict could be equally revealing. Sadly, there is no shortage of historical and contemporary events to explore.

References

Baudendistel, Rainer. 2006. *Between bombs and good intentions: The red cross and the Italo-Ethiopian war, 1935–1936*. Oxford: Berghahn Books.

Bouchet-Saulnier, Françoise. 2015. *The practical guide to humanitarian law*. https://guide-humanitarian-law.org/content/article/3/humanitarian-principles/.

Convention (IV) relative to the Protection of Civilian Persons in Time of War, Geneva, 12 August 1949, Articles 18 and 19.

De Baets, Antoon. 2022. The view of the past in international humanitarian law (1860–2020). *International Review of the Red Cross*. 104 (920–921, November): 1586–1620. https://international-review.icrc.org/articles/the-view-of-the-past-in-ihl-920.

Hofman, Michiel. 2021. Naming and shaming the bombers. In *Everybody's war: The politics of aid in the Syria crisis*, ed. Jehan Bseiso, Michiel Hofman, and Jonathan Whittall, 185–210. Oxford: Oxford University Press.

Jessup, Philip C., and Francis Déak. 1935. *Neutrality: Its history, economics and law in four volumes, I: The origins*. New York: Octagon Books.

Langendorf, Jean-Jacques. 2007. *Histoire de la Neutralité : Une perspective*. Gollion: Infolio éditions.

Le conflit Italo-Ethiopien et la Croix-Rouge. 1936. Rapport du Comité International de la Croix-Rouge. December.

Letters from Overseas Council, Church of Scotland to ICRC, 22 February and 26 March 1968. ICRC archives.

Lieber, Francis. 1863 (1898). *Instructions for the government of armies of the United States in the field*. Washington, DC: Government Printing Office.

McGreal, Stephen. 2008. *The war on hospital ships: 1914–1918*. Barnsley: Pen & Sword Maritime.

McLean, Duncan. 2019. Medical care in armed conflict: Perpetrator discourse in historical perspective. *International Review of the Red Cross*. 101 (911, April): 771–803. https://library.icrc.org/library/docs/DOC/irrc-911-mclean.pdf.

Mendiharat, Pierre. 2017. Hospital airstrikes: Gathering evidence through images. Interview by Agnes Varraine-Leca, Médecins Sans Frontières. February 15. https://msf-crash.org/en/blog/war-and-humanitarianism/hospital-airstrikes-gathering-evidence-through-images

Minear, Larry. 1999. The theory and practice of neutrality: Some thoughts on the tensions. *International Review of the Red Cross*. 1 (833, March): 63–71. https://international-review.icrc.org/sites/default/files/S156077550009235Xa.pdf.

Médecins Sans Frontières. 2014. *MSF hospital bombed in Sudan*, June 17. https://www.msf.org/msf-hospital-bombed-sudan.
———. 2015a. *MSF hospital bombed in South Kordofan*, January 22. https://www.msf.org/sudan-msf-hospital-bombed-south-kordofan.
———. 2015b. *Initial MSF internal review: Attack on Kunduz Trauma Centre, Afghanistan*. November. https://www.msf.org/afghanistan-msf-releases-internal-review-kunduz-hospital-attack.
———. 2016a. MSF internal review of the January 2016 attack on Shiara hospital. Internal Document, March 15. https://arhp.msf.es/resource/msf-internal-review-of-the-january-2016-attack-on-shiara-hospital/.
———. 2016b. MSF internal review of the 15 August attack on Abs hospital Yemen. Internal Document, September 27. https://arhp.msf.es/wp-content/uploads/2023/07/Yemen_Abs_investigation_summary_final_0-3.pdf.
———. 2016c. Minimum standards for MSF internal reviews of attacks against MSF facilities, personnel and patients. Internal Document, December.
———. 2021a. Analysing the unthinkable—One year after the Dasht-e-Barchi attack. May 10. https://www.msf.org/one-year-after-attack-dasht-e-barchi-afghanistan.
———. 2021b. Attacks on medical care: Syria, South Sudan, Yemen, Afghanistan, and more, health facilities have been attacked, looted and destroyed. June 8.
Müller, Leos. 2019. *Neutrality in world history*. London: Routledge.
Protocol additional to the Geneva conventions of 12 August 1949, and relating to the protection of victims of international armed conflicts (protocol I), Geneva, 8 June 1977, Articles 12–13.
Protocol additional to the Geneva conventions of 12 August 1949, and relating to the protection of victims of non-international armed conflicts (protocol II), Geneva, 8 June 1977, Article 11.
Rubenstein, Leonard. 2021. *Perilous medicine: The struggle to protect health care from the violence of war*. New York: Columbia University Press.
Unknown Author. 1918. III.—Diplomatic correspondence: Memorandum of the German government respecting the misuse of enemy hospital ships. In *The war on hospital ships: With narratives of eye-witnesses and British and German diplomatic correspondence*. London/New York: Harper & Brothers Publishers.
———. 1937. Internal update to Geneva. October, 21. ICRC archives.
Wani, Kentaro. 2017. *Neutrality in international law: From the sixteenth century to 1945*. London: Routledge.
Weapons Law Encyclopaedia. 2013. https://www.weaponslaw.org/glossary/barrel-bomb.

Chapter 3
Dual Loyalties in Military Medicine. Some Lessons from the Past—Challenges for the Future

Dilemmas and Responsibilities of Health Professionals in Relation to Human Rights, in Particular of Military Medical Personnel

Adriaan van Es

Abstract Physicians have sworn the oath of Hippocrates and are bound in loyalty to their patients. But they may also face loyalty to their employer, government or commanding officer. Such dual loyalty problems exist inter alia in the armed forces and can lead to ethical dilemmas.

This chapter describes dual loyalty problems of health professionals in the context of a human rights-based approach, describes medical involvement and complicity in violations of human rights, but also the mobilization of medical knowledge in the defense and promotion of human rights. History and cases from the past can provide lessons, and challenges for the future.

Focusing on military health professionals, the chapter concludes that the military health professional's first and overruling identity and priority is that of a health professional, positioned in the medical community at large—on individual and organizational basis—and should be facilitated to report and expose violations of international human rights law and international humanitarian law.

The author stresses that physicians and other health professionals should never remain silent about human rights violations and always defend patient rights and professional medical ethics when they witness human rights abuses, in general and in the armed forces.

A. van Es (✉)
Amersfoort, The Netherlands

D. Messelken, A. E. Barbar (eds.), *Challenging Medical Neutrality*, Military and Humanitarian Health Ethics, https://doi.org/10.1007/978-3-031-69398-4_3

1 Introduction

The role and position of physicians in the armed forces has been debated since long time by professionals of law, the military, health, philosophers and others. The question of "physician first, soldier next" or vice-versa, handling of dual loyalty situations (loyalty to the—wounded or sick—patient or to the commanding officer, employer or (even) court verdict) has never lost its relevance, urgency and topicality.

It is therefore interesting and relevant to draw lessons from the past, since they may offer challenges for the future. The way in which professionals have handled dual loyalties in the past and how professional organizations and human rights bodies have discussed ensuing dilemmas are important elements to take into account when discussing military medical ethics and *Medical Neutrality in Theory and Practice*, as was the title of the 12th International Committee of Military Medicine's (ICMM) workshop on Military Medical Ethics.

The debate about theory and practice, ethics, philosophical considerations, and historical research of medical neutrality, especially in military medicine has a very long history. In this chapter I will discuss these issues from a human rights-based perspective, from my own experience as physician and as human rights advocate, and with the conviction that military medicine is part of medicine at large and that military health professionals are members of the health professional community. I shall discuss dual loyalty (also called "mixed agency") issues in broader context than of military medicine alone. And, as much of the debate has a long history, I will present cases, histories, experiences and discussions that illustrate the importance of lessons from the past and the importance of remembrance.

Human rights and the role of health professionals can be divided in two broad areas of concerns that need study, analysis and action. The first concern is the broad field of involvement of health professionals in of medical procedures or quasi-medical procedures and activities that violate human rights. These involvements may be legally required, happen under duress or take place voluntarily. Examples are corporal punishment and virginity testing. The—opposing—second concern is the application of medical expertise in promoting and/or defending human rights. Physicians and nurses are often the first, sometimes the only ones to witness gross human rights violations. Both on individual and organizational level health professionals can play a pivotal role in the defense of human rights. Examples are adequate independent forensic examination, providing medical expert opinion (witness) in court, developing procedures and protocols for adequately describing (sequelae of) human rights violations—such as the Istanbul Protocol—and defending colleagues who refuse to be or get involved in human rights violations.

2 Corporal Punishment

Very early in my career I encountered a challenging situation of dual loyalty: in the district hospital in Ghana where I worked in the late 70's, I was regularly requested to certify the fitness for corporal punishment of a convicted person. This heritage of

British colonial rule—since many years abandoned in Ghana, but still practiced in 28 countries[1]—meant that the offender had to undergo corporal punishment by caning, but he had to be mentally and physically fit to undergo that punishment, and the person should not be a minor. This had to be certified by a physician. This legal obligation implied that I would become involved in a system where the physical and mental integrity was violated. Besides the lack of scientific or clinical basis to assess someone's age with a "clinical eye", this involvement of a physician in the juridical system that allowed (or prescribed) violation of physical integrity by a method that in principle does not differ from torture, implied a violation of the physician's Hippocratic oath.

After a few cases I refused to participate in this practice. A quick survey among Dutch tropical doctors in Africa, learned that many of them were supposed to provide such certificates of fitness for punishment. After the survey the Dutch Ministry of Foreign Affairs issued a letter that Dutch doctors were not supposed to cooperate in this practice.

The dilemma of this dual loyalty choice could be compounded by the convict's preference (and urge to be certified fit for punishment) to undergo the caning rather than the alternative of imprisonment in inhuman conditions. Compounding considerations like this also play an important role in discussions about medical involvement, and thus complicity in the death penalty, torture and forced feeding in hunger strikes. However, the *UN Convention against Torture and Other Cruel, Inhuman or Degrading Treatment or Punishment* of 1984 (United Nations 1984) and the World Medical Association *Declaration of Tokyo* 1975 (World Medical Association (WMA) 1975) explicitly prohibit any involvement of physicians in torture, cruel and inhuman treatment, including forced-feeding of persons on hunger strikes.

The debate about medical involvement in corporal punishment continues, as illustrates the comment *The ethics of penal amputation* in the Indian Journal of Medical Ethics (2020). A professor in surgery in Malaysia, proposes an ethical basis for the involvement of a surgeon in the Hadud (Sharia) amputation of a limb, when the amputation is done with the free consent of the convicted criminal. He compares the situation with physicians' participation in capital punishment and with euthanasia. He argues that "doctors have an ethical obligation to reduce the suffering of, say, a dying cancer patient. A corollary, therefore, is that the dying condemned prisoner also deserves the best 'treatment' that the doctor can give" (Das 2020). In the case of penal amputation, for example, the prisoner deserves a surgeon's specialized services rather than facing an amputation by untrained person.

Upon invitation of the editors of the Journal I wrote a discussion comment *The crime of medical complicity in penal amputation*. Remembering that "this dilemma has a very long history and the number of medical 'humanization' of cruel forms of punishment is equally long", that "the dilemma and its often horrific consequences have been investigated and exposed by human rights and professional organizations, always leading to moral, legal and professional condemnation", that "the

[1] https://en.wikipedia.org/wiki/Judicial_corporal_punishment

sanctity of medical ethics and the prohibition against inflicting harm is reflected in many professional codes of conduct and human rights conventions, including the Islamic Code of Medical Ethics" (Van Es 2020). I quoted the International Conference on Islamic Medicine in Kuwait in 1981, which resulted in the *Islamic Code of Medical Ethics*, which states: "The medical profession shall not permit its technical, scientific, or other resources to be utilized in any sort of harm or destruction or infliction upon man of physical, psychological, moral, or other damage … regardless of all political or military considerations" (Islamic Organization for Medical Sciences 1981). I also quoted The World Medical Association *Declaration of Tokyo* on torture and cruel, inhuman and degrading treatment or punishment that states: "The physician shall not countenance, condone or participate in the practice of torture or other forms of cruel, inhuman or degrading procedures, whatever the offense of which the victim of such procedures is suspected, accused or guilty" (World Medical Association (WMA) 1975); and I stressed that the Malaysian Medical Association is a member of the World Medical Association.

As physicians may be pressured to participate in corporal punishment, strong and independent professional medical organizations may support efforts in upholding medical ethics and opposing malpractice and complicity in ill-treatment. Two important examples of these role in providing medical professional opposition to human rights violations are the British Medical Associations *Medicine Betrayed* report (British Medical Association 1992) and the Turkish Medical Association 1984 ban on participation of its members in capital publishment.

3 Health and Human Rights Advocacy

The experience with corporal punishment triggered my interest and decades of involvement in health-related human rights advocacy. In the mid-80's of the last century health and human rights organizations were established, such as *Physicians for Human Rights* (PHR) and in the Netherlands the *Johannes Wier Foundation* (Johannes Wier was a sixteenth century Dutch physician who opposed effectively the hunting and execution of "witches"). The groups focused on a large variety of subjects and issues, in publications, projects and conferences. A selected—not exhaustive—number of examples:

3.1 Opposition Against Physician Involvement in the Death Penalty and Corporal Punishment

In 1994 PHR together with other institutions published the report *Breach of Trust* that "documents that physicians continue to be involved in executions, in violation of ethical and professional codes of conduct. This involvement is often mandated by state law and specified in departmental regulations about execution procedures", and emphasizes that "execution is not a medical procedure, and is not within the scope of medical practice", and recommends i.e. that "U.S. laws do not require physicians to violate professional ethics" (American College of Physicians et al. 1994). Later in 2013, the WMA *Resolution on Physician Participation in Capital Punishment* reinforced the unethical aspects of participation of physicians in capital punishment (World Medical Association (WMA) 2013).

3.2 Opposition Against Physician Involvement and Complicity in Torture and Cruel and Inhuman Treatment or Punishment

Examples of medical involvement and complicity in torture and abuse of medical knowledge and practice exist in many countries, including in the world's major powers USA and China. In the report *Break them down*, PHR describes acts and omissions of health professionals involved in state intelligence, who violated their professional ethics, undermining trust between patients and doctors and breaking the law (Physicians for Human Rights PHR 2014). In 2022 the United Nations (UN) Office of the High Commissioner for Human Rights published the report *Assessment of human rights concerns in the Xinjiang Uyghur Autonomous Region, People's Republic of China: Forced abortions and sterilizations* (Office of the High Commissioner for Human Rights 2022a) stating serious concern about reported instances of the use of coercive measures, including forced abortion and forced sterilization, with a view to limiting births. In a press release UN human rights experts said they were alarmed by reports of alleged "organ harvesting" targeting minorities, including Uyghurs, Tibetans, Muslims and Christians, in detention in China.[2] On a particular resolution addressing the Chinese case and alleged "organ donations" by prisoners, the World Medical Association (WMA) notes that "prisoners and other individuals in custody are not in a position to give consent freely, and therefore their organs must not be used for transplantation", pointing to the malpractice that entails the involvement of doctors in such actions (World Medical Association 2006).

3.3 Opposition Against Torture in Healthcare Settings

The campaign *Stop Torture in Healthcare*, launched in 2011, focused on coerced sterilization and access to adequate pain treatment.[3] Cases of coerced and forced sterilization have been reported in countries across the globe. Persons have been

[2] https://www.ohchr.org/en/press-releases/2021/06/china-un-human-rights-experts-alarmed-organ-harvesting-allegations

[3] https://www.opensocietyfoundations.org/voices/stop-torture-health-care-0

forcibly sterilized without knowing the procedure had been performed or without been given the opportunity to consent. Others were coerced into sterilization by financial or other incentives, denial of medical services, or by being misinformed or otherwise compelled to undergo the procedure. Practices of forced and coerced sterilizations performed on women living with HIV/AIDS have been documented in certain South-American and African States. In Europe cases of forced and coerced sterilization have been documented against women of ethnic minorities. In 2005, a Czech Ombudsman issued a report in which he reviewed more than 80 allegations of forced and coerced sterilization of women, most of whom are members of the Roma minority (IFHHRO 2011).

On denial of access to adequate pain treatment, a report to the Human Rights Council by the UN Special Rapporteur on Torture and other Cruel, Inhuman or Degrading Treatment or Punishment noted that: "the de facto denial of access to pain relief, if it causes severe pain and suffering, constitutes cruel, inhuman or degrading treatment or punishment" and that "all measures should be taken to … overcome current regulatory, educational and attitudinal obstacles to ensure full access to palliative care" (UN Human Rights Council 2009). The Medical Human Rights Network (previously known as the International Federation of Health and Human Rights Organisations) organized in 2011 an international conference of pain and palliative care organizations, health professional organizations and human rights organizations to research and discuss the denial of access to adequate pain treatment worldwide, resulting in a resolution of the WMA on the Access to Adequate Pain Treatment (World Medical Association (WMA) 2011). This resolution insists that access to adequate pain treatment is a human right and that health care workers must offer pain assessment and pain treatment to patients with pain.

3.4 *Opposition Against Quasi 'Medical' Procedures, Which in Fact Are Forms of Torture or Cruel and Inhuman Treatment or Punishment*

When addressing the issues of vaginal testing to assess virginity, the International Forensic Expert Group (IFEG) concludes that "virginity examinations are medically unreliable and inherently discriminatory, and, in almost all instances, when conducted forcibly, result in significant physical and mental pain and suffering" (Independent Forensic Expert Group 2015). They also point out that such procedures have no clinical or scientific value and constitute cruel, inhuman and degrading treatment, potentially amounting to torture depending on the individual circumstances. IFEG goes on to insist that when virginity examinations are forcibly conducted, involving vaginal penetration, such procedures should be considered a form of sexual assault and rape. Thus any involvement of the health professional in these examinations is violating basic ethical principles and constitute mal practice.

In the same line, the IFEG has delibered that forcibly conducted anal examinations have no medical or scientific value in determining whether consensual anal intercourse has taken place. Moreover, these examinations are inherently discriminatory and, in almost all instances, result in significant physical and mental pain and suffering, representing cruel, inhuman, and degrading treatment, any may even be understood as torture depending on the individual circumstances. Similarly to vaginal virginity testing, forcibly conducted anal penetration under such excuses may be considered a form of sexual assault and rape, therefore health professionals should refrain from participating in such practices.

Sexual identity and orientation are neither a disease nor a crime. Health professionals, therefore, have no role in diagnosing it or aiding State officials in policing and punishing people on the basis of their sexuality through any means of testing or examination. Health professionals who conduct anal examinations or other tests targeting "male homosexuality" are knowingly or unknowingly perpetuating social customs and norms that violate human rights and human dignity and are ultimately facilitating and participating in cruel, inhuman, and degrading treatment, sexual assault, and possibly torture. National medical associations should take action to unequivocally ban these practices, hold practitioners accountable, and work with civil society and government officials to end laws that criminalize sexual identity and orientation (Independent Forensic Expert Group 2016).

In light of such, conversion therapies were also identified by the IFEG as having no medical or scientific validity, representing an ineffective, inherently repressive practice. Such therapies may cause individuals significant or severe physical and mental pain and suffering with long-term harmful effects and as a form of cruel, inhuman, or degrading treatment or torture, "states have an obligation to ensure that both public and private actors are not directly committing, instigating, inciting, encouraging, acquiescing in or otherwise participating or being complicit in conversion therapy". The IFEG goes on insisting that" health professionals should refuse to conduct conversion therapy and report colleagues who advertise, offer, or perform them to the appropriate authorities, (...) national medical and mental health associations should take steps to hold practitioners accountable and work with civil society and government officials to pass laws that ban conversion therapy" (Independent Forensic Expert Group 2018).

4 Mobilization and Development of Medical Expertise in the Protection of Human Rights

In many fields the introduction and mobilization of medical knowledge and expertise in the defense and promotion of human rights has proven to be effective, essential and pivotal. A clear example is the development of the *Istanbul Protocol, Manual on the Effective Investigation and Documentation of Torture and Other Cruel, Inhuman or Degrading Treatment or Punishment*, first published 1999, adopted by the UN in 2004 and updated in 2022. The Istanbul Protocol is the "internationally agreed set of best-practice standards that States must follow in investigating torture and by which health and legal professionals should evaluate allegations of torture, document the cases of survivors and produce quality evidence" (Office of the High Commissioner for Human Rights 2022b).

A medical expert opinion can be crucial in defending human rights, as was demonstrated in 2014, when the Grand Chamber of the European Court of Human Rights delivered a ruling in the Câmpeanu versus Romania case. Câmpeanu, a Roma boy with mental disability and HIV due to a childhood blood transfusion, had died as a result of gross medical negligence. The Court ruled that the Romanian government had breached his right to life, and criticized the authorities for not conducting any investigation (European Court of Human Rights 2014).

As early as 1996, Professor Jack Geiger flagged the importance of mobilizing public health expertise to reduce violations of human rights and provides a list of epidemiological studies and public health techniques in defense of medical neutrality. These include, amongst others, interviewing and documenting situations that may related or directly amount to violations of human rights and international law; making ethical use of medical data of victims of alleged human rights abuses; performing autopsies and other forensic enquiries on victims of suspected torture, extrajudicial death, deaths in detention and other complex scenarios such as bodies found in mass graves or unmarked individual graves; documenting health effects of indiscriminate weapons and effects of weapons of mass destruction; supporting efforts for the health and well-being of prisoners; conducting epidemiological analysis of effects of war and deprivation of basic rights to sustain the defense of human rights; supporting, protecting and defending health workers detained, abused or tortured; investigate cases of violation of ethics by health workers (Geiger 1996).

In his book *Perilous Medicine, the struggle to protect health care from the violence of war* Leonard Rubenstein describes principles designed and meant to safeguard medical practice in wartime, and the ongoing failure to respect them (Rubenstein 2021). The book documents in great detail how violent targeting of health workers has become a familiar fact of modern warfare. This fact calls urgently for policies that keep up to the effective defense of protected status of medical premises, personnel and transports. Concluding his book, Rubenstein calls for more and coordinated health activism. He rejects a cynical acceptance of the seemingly "new normal" with its lack of respect for international standards. He highlights signs of hope, such as political mobilization against arms sales in many countries including in producer countries such as the UK and USA.

The Safeguarding Health in Conflict Coalition provides regular reports with information on attacks against healthcare facilities in personnel in different countries, also calling for further protection and involvement of the medical community in the defense of healthcare in conflict-affected settings.

5 Military Medicine: First Doctor, Then Soldier

An important question in the medical and military communities is no doubt: as a military doctor, are you first doctor or soldier? An interesting recent article and interview in the Dutch newspaper *Trouw*[4] illustrates this discussion. The article shows two photos of a surgeon and an intensivist, in military uniform, and, on another photo, in civilian hospital gowns. The physicians express their allegiances in the interview: first soldier, then doctor. They ascertain that the dilemmas are rare and although in disagreement with the author of this article, they express that their first allegiance is to the military.

The WMA *Regulations in times of armed conflict and other situations of violence* (adopted 1956, revised 1957, 1983, 2004, 2012 and 2012) are quite clear about the position of the physician in the armed forces: "medical ethics in times of armed conflict is identical to medical ethics in times of peace, as stated in the International Code of Medical Ethics of the WMA. If, in performing their professional duty, physicians have conflicting loyalties, their primary obligation is to their patients; in all their professional activities, physicians should adhere to international conventions on human rights, international humanitarian law and WMA declarations on medical ethics" (World Medical Association (WMA) 2012).

I view military doctors as medical colleagues, and I believe they should be part of the professional community of physicians. In the Netherlands there is no military medicine chapter in the Royal Dutch Medical Association. We have chapters of GP's, specialists, PH specialists, but no chapter of military doctors. I think this is also the case in many other countries Why is this important? It is important that military doctors belong in mainstream medicine and are members of the medical community at large, to stress and promote a sense of collegiality, communication and debate about all kind of fields, including ethics and professional behavior. And to help protect military doctors to not deviate from the oath they have pledged one day. Wouldn't it be a good idea to put membership of military medical organizations in the realm of national medical associations on the agenda of the WMA and the ICMM?

History reminds us of the importance of the principle 'physician first'. In the Nuremberg doctor's trial of 1947 the most horrific misconduct of doctors was exposed to the world, and the international medical community was shaken to its core, becoming aware that medical ethics had been violated so deeply. In response the World Medical Association was established, and the Hippocratic oath was revived into the Declaration of Geneva. But this crucial lesson was not the last one, many more would come, such as the violations of medical ethics under the racist apartheid regime in South Africa. I visited apartheid South Africa in 1987, during the state of emergency, to attend a conference of anti-apartheid health professionals (National Medical and Dental Association). These health professionals had to pay

[4] https://krant.trouw.nl/titles/trouw/8321/publications/1759/pages/30

their adherence to medical ethics with threats, imprisonment, exclusion from organizations and hospital jobs, even with torture and death.

After the fall of the apartheid regime, the Truth and Reconciliation Committee held hearings of the health sector. These hearings exposed an incredible number of cases of discrimination, torture, death, structural denial of access to health care and involvement of doctors in gross human rights violations (Baldwin-Ragaven et al. 1999). In the aftermath of these hearings the Dual Loyalty project was launched: a multidisciplinary study, conference and publication of dual loyalty situations in which doctors, nurses and other health professionals can be squeezed between their professional ethical obligation to their patients on the one side, and their real or perceived obligation toward their employer, government or commanding officer. The project was initiated by PHR and the University of Cape Town. An International Working Group of 56 experts from 14 countries identified a large number of situations where dual loyalties become apparent and formulated guidelines for specific groups where health professionals are likely to meet dual loyalty problems: prison and other custodial situations, healthcare for immigrants and asylum seekers, health professionals in the workplace, health professionals engaged in forensic evaluations and military health professionals. The report *Dual Loyalty & Human Rights In Health Professional Practice; Proposed Guidelines & Institutional Mechanisms* identifies types of dual loyalty practices that violate human rights, including many of the cases and situations previously listed in this article. It also includes disclosing confidential information in circumstances that violate human rights, and remaining silent in face of violations of human rights (Physicians for Human Rights 2002).

In the paper *Dual Loyalty among Military Health Professionals: Human Rights and Ethics in Times of Armed Conflict*, based on the Dual Loyalty project report, me and fellow co-authors state that "dual loyalty conflicts are […] common in the military, occurring in a variety of situations. Involvement of military medical personnel in torture whereby medical skills have been used to inflict pain or physical or psychological harm on an individual that is not a legitimate part of medical treatment has been documented all too frequently" (London et al. 2006). While international guidelines for physicians working in military services exist, any dual loyalty situation may render a physician vulnerable to pressure to conform to state expectations, therefore guidelines alone will be meaningless unless coupled with strategies to support health professionals and address the systemic forces that affect how medicine is practiced in a particular setting. A key structural element is the need for military organizations to recognize the obligations medical personnel have, so that they are not forced into disobeying an order on ethical grounds.

It is also important to state that, if and when human rights concerns arise, placing responsibility for its resolution on the health professional is untenable. Decision-making should be devolved to a multidisciplinary medical ethical structure with active civilian participation and oversight, such as a commission with membership that includes an adequate number of civilian health professionals skilled in ethical issues and human rights, providing the needed balance in determining what kind of military necessity justifies deviating from the norms of ethical medical practice.

An example of civilian oversight did take place in the Netherlands on the Srebrenica case, as is described by Siemons in Medisch Contact in 1996: the Health Care Inspectorate launched an investigation aiming to determine whether members of the military medical unit failed to provide medical care for civilian casualties and to assess whether towards individual members of the team (disciplinary) measures be taken. The Health Care Inspectorate criticizes the team for not providing first aid in a traffic accident, a surgeon was criticized for not providing aid to an injured civilian woman, after a summary triage. However, the Health Care Inspectorate found the imminent threat of war and often confusing conditions mitigating factors, and decided the case should not be submitted to the medical disciplinary board (Siemons 1996).

Accountability was demonstrated by Kremer, surgeon and colonel in the Dutch army. Reflecting on the failures of Dutchbat, he wrote: "In this chaotic, threatening and frightening situation, the self-evident humanitarian aid has not always been so self-evident. Opposing opinions of military specialists, the command structures that are not always clear and the very difficult circumstances certainly contributed to this. The age-old debate about the compatibility between the military oath and the deepest sense of human culture brought to the fore [...]. The law of war is regulated in international treaties, such as the Geneva Conventions of 1949 and the additional protocols of 1977, ratified by the Netherlands in 1954 and 1987 respectively. [...] Reserving medical supplies exclusively for (Dutch) military personnel should therefore not have been allowed. Nor was it right to refuse adequate medical assistance to a civilian whose life was in danger. Refusal by the deputy commander to take over seriously injured civilian patients without consultation with those with final medical responsibility was also incorrect. When deploying military doctors to war zones, more attention should be paid to the dilemma of possible conflicting military, medical and humanitarian interests. In drawing up the humanitarian policy, both the military doctors and the military commanders must have their input, so that there are no more misunderstandings about this. Better explanation of the international treaties on human rights is necessary. An improvement of the regulations regarding the job description of military doctors has already been implemented thanks to the 'lessons learned in Yugoslavia'. Humanitarian assistance in acute situations will become mandatory; the military doctors must adhere to the doctor's oath they have taken. Continuous evaluation and improvement of medical services and the maintenance of medical neutrality remain necessary" (Kremer 1996 translated by the author).

Guidelines stress that military doctors should report violations of human rights that interfere with their ability to comply with their duty of loyalty to patients. It is well known that it is quite difficult to be a whistleblower; still this is much needed, and whistleblower protection measures should be in place. Although they are the first (and often only) ones to witness severe human rights violations, health professionals are not among the whistleblowing frontrunners. This applies as well to military health professionals. A Dutch military historian recently noted that only very seldom health professionals report violations of the Geneva Conventions. He wrote me, in a personal communication: "my focus was indeed on the question of why you encounter few whistleblowers in the military-medical world, even though the abuses are often serious. Presumably this also has to do with the 'split' loyalty to both the Hippocratic oath and the military organization. I suspect that similar processes are at play in the civil-medical world: loyalty to colleagues, organization."

6 Conclusion

Physicians and other health professionals should never remain silent about human rights violations and always defend patient rights and professional medical ethics when they witness human rights abuses, in general and in the armed forces.

The military health professional's first and overruling identity and priority is that of a health professional, positioned in the medical community at large—on individual and organizational basis—and should be facilitated to report and expose violations of international human rights law and international humanitarian law.

References

American College of Physicians, Human Rights Watch (Organization), National Coalition to Abolish the Death Penalty (U.S.), and Physicians for Human Rights (U.S.), ed. 1994. *Breach of trust: Physician participation in executions in the United States*. Philadelphia/New York/Washington, DC/Boston: American College of Physicians ; Human Rights Watch ; National Coalition to Abolish the Death Penalty ; Physicians for Human Rights.

Baldwin-Ragaven, Laurel, Jeanelle De Gruchy, and Leslie London, eds. 1999. *An ambulance of the wrong colour: Health professionals, human rights and ethics in South Africa*. Rondebosch, SA: University of Cape Town Press.

British Medical Association. 1992. *Medicine betrayed : The participation of doctors in human rights abuses*. London/Atlantic Highlands: Zed Books.

Das, Anjan K. 2020. The ethics of penal amputation. *Indian Journal of Medical Ethics* V: 143–148. https://doi.org/10.20529/IJME.2020.032.

European Court of Human Rights. 2014. Case of Centre for legal resources on behalf of Valentin Câmpeanu V. Romania.

Geiger, H. Jack. 1996. Public health expertise in defense of medical neutrality. *Medisch Contact* 51: 1460–1461.

IFHHRO. 2011. *Forced and / or coerced sterilisation*. Utrecht: IFHHRO.

Independent Forensic Expert Group. 2015. Statement on Virginity Testing. *Torture Journal* 25: 7. https://doi.org/10.7146/torture.v25i1.109509.

———. 2016. Anal examinations in cases of alleged homosexuality. *Torture Journal* 26: 7. https://doi.org/10.7146/torture.v26i2.108205.

———. 2018. *Statement on conversion therapy*. IFEG.

Islamic Organization for Medical Sciences. 1981. *Islamic Code of Medical Ethics Kuwait Document*.

Kremer, G.D. 1996. Medische neutraliteit in crisisbeheersingsoperaties. Hoezo vanzelfsprekend? *Medisch Contact* 51: 1462–1464.

London, Leslie, Leonard S. Rubenstein, Laurel Baldwin-Ragaven, and Adriaan Van Es. 2006. Dual loyalty among military health professionals: Human rights and ethics in times of armed conflict. *Cambridge Quarterly of Healthcare Ethics* 15: 381–391. https://doi.org/10.1017/S096318010606049X.

Office of the High Commissioner for Human Rights. 2022a. *OHCHR assessment of human rights concerns in the Xinjiang Uyghur autonomous region, People's Republic of China*. UNO.

———. 2022b. *Istanbul protocol. Manual on the effective investigation and documentation of torture and other cruel, inhuman or degrading treatment or punishment*. New York: UNO.

Physicians for Human Rights. 2002. *Dual loyalty & human rights in health professional practice: Proposed guidelines & institutional mechanisms*. Boston: Physicians for Human Rights.

Physicians for Human Rights PHR. 2014. *Doing harm: Health professionals' central role in the CIA torture program*. PHR.

Rubenstein, Leonard S. 2021. *Perilous medicine: The struggle to protect health care from the violence of war*. New York: Columbia University Press.

Siemons, G.H.A. 1996. Medische hulp aan burgerslachtoffers. Srebrenica onder de loep. *Medisch Contact* 51: 1465.

UN Human Rights Council. 2009. *Promotion and protection of all human rights, civil, political, economic, social and cultural rights, including the right to development. Report of the special rapporteur on torture and other cruel, inhuman or degrading treatment or punishment*. New York: UNO.

United Nations. 1984. *Convention against torture and other cruel, inhuman or degrading treatment or punishment*. New York: UNO.

Van Es, Adriaan. 2020. The crime of medical complicity in penal amputation. *Indian Journal of Medical Ethics* V: 149–151. https://doi.org/10.20529/IJME.2020.034.

World Medical Association. 2006. WMA council resolution on organ donation in prisoners. *WMA*.

World Medical Association (WMA). 1975. *WMA declaration of Tokyo—Guidelines for physicians concerning torture and other cruel, inhuman or degrading treatment or punishment in relation to detention and imprisonment*. WMA.

———. 2011. *WMA resolution on the access to adequate pain treatment*. WMA.

———. 2012. *WMA regulations in times of armed conflict and other situations of violence*. Geneva: WMA.

———. 2013. *WMA statement on the United Nations resolution for a moratorium on the use of the death penalty*. WMA.

Part II
Medical Neutrality in Theory

Chapter 4
Military Health Care Personnel's Moral Stance on War: Medical Neutrality and Humanity

Eva van Baarle

Abstract This chapter focusses on the moral value humanity underlying the principle of medical neutrality (also referred to as political neutrality). Should it be considered an obligation, as being unconditional and in that sense, an absolute moral value? It seems that both the absolute and the relative are at play here. The contradictory logic in humanity appears similar to the paradox Jacques Derrida identified within the concept of forgiveness. There is the unconditional value of humanity, like charity, to provide medical care to those in need. At the same time, there is the pragmatic imperative of historical, legal or political conditions which demand the opposite (taking sides). These two remain irreducible: medical action thus has to be related to a moment of unconditionality to avoid being reduced to a strategic, political or economic calculation. On the other hand, such unconditionality can hardly be permitted in operational practice, as decisions would be deduced from incontestable ethical precepts or principles. It requires respect for both extremes. Military health care personnel are thus continually confronted with immensely difficult decisions: whether to judge, or condemn, who to deem innocent and where to show humanity and treat the sick and wounded.

1 Introduction

The principle of medical neutrality is grounded in International Humanitarian Law as well as in Human Rights Law, implying that medical personnel caring for the wounded require immunity and protection. In order to be protected, medical personnel should have no concerns about the moral status of those they are offering care, irrespective of who they are. In days gone by, when volunteer medical personnel often worked closely with military medical health care personnel, they both required

E. van Baarle (✉)
Faculty of Military Sciences, Netherlands Defence Academy, Breda, The Netherlands

Amsterdam UMC, Location VUmc, VU University Amsterdam, Amsterdam, The Netherlands
e-mail: em.v.baarle@mindef.nl

D. Messelken, A. E. Barbar (eds.), *Challenging Medical Neutrality*, Military and Humanitarian Health Ethics, https://doi.org/10.1007/978-3-031-69398-4_4

and eventually came to enjoy identical protection (Gross 2007; Rubenstein and Bittle 2010). The principle of neutrality thus passed seamlessly from volunteers to military health care personnel. By the twentieth century, volunteers were no longer needed when military organizations expanded their medical personnel. The volunteers may have gone, yet the principle of medical neutrality they initiated remained intact. Today, medical neutrality is often construed as a professional or moral *obligation* or *duty*, requiring military health care personnel to treat the wounded and injured impartially, whether friend or foe, based on medical need and urgency (List 2008; Benton and Atshan 2016).

This chapter focusses on understanding military personnel's medical neutrality and the underlying moral value humanity. Should it be seen as an obligation, as being unconditional and, in that sense used as an absolute moral value, or considered a relative or instrumental value? Inspired by French philosopher Jacques Derrida's work on ethical themes like forgiveness (Derrida 2001), this chapter exposes and reflects on some of the binary opposites that underpin dominant thinking about the principle of medical neutrality and the meaning of humanity for military health care personnel. Within the moral value humanity, a contradictory logic may be at play, similar to the paradox Derrida identified within the concept of forgiveness (Wortel 2009). His work leads us into a discussion about the moral value of humanity and the implications for military health care professionals' moral stance on war.

1.1 On Forgiveness

French philosopher Jacques Derrida is best known for his work on deconstruction, a philosophical approach seeking to uncover underlying assumptions and contradictions in language and thought. In *On Cosmopolitism and Forgiveness* (Derrida 2003) he argues that forgiveness confronts us with a paradox: Small things that are forgivable do not really seek forgiveness, they are easy to forgive. If something is easy to forgive, does it really demand forgiveness? Even a crime that is potentially forgivable may not need forgiveness. Derrida asks us to think about what it really means to forgive. The notion of forgiveness stems from Jewish and Christian traditions. Although Christianity recommends or even grants everyone forgiveness, it nevertheless seemingly stands for something that is not humanly possible: how can we remember the evil committed, and at the same time not blame the one who committed it? (Van Tongeren 2008). If there is something to forgive, it would be the worst unforgivable crime or harm.

Drawing on this paradox, Derrida examines the work of Vladimir Jankélévitch, who speaks about a duty of non-forgiveness. According to Jankélévitch forgiveness died in the death camps: The terrible acts committed in the camps by Nazi Germany are irreparable, therefore they are unforgivable (Jankélévitch 1965). Derrida disagrees, arguing that the notion of forgiveness, including an imperative of love, contains an unresolved and unresolvable paradox between unconditional and conditional forgiveness. Unconditional, implying gracious, infinite forgiveness, granted to the guilty and even to those who do not ask for forgiveness, is not possible: "This forgiveness is mad...a madness of the impossible" (Derrida 2003, 39). Conditional forgiveness, proportional and carefully considered, granted to the regretful, implies they are no longer guilty in the same way as before. True forgiveness that is worthy of the name can never *only* be conditional forgiveness; nor is it absolutely unconditional. Its in-between 'space' is irreducible, always caught up in the tension between these two extremes. Deliberation and deciding on forgiveness thus take place in this tensed in-between. We need to understand what we are doing when we forgive: understand that forgiveness is an engagement that includes an element of unconditional forgiveness that we must risk, but that we can never ultimately justify.

In the following section I contend that there is a similar paradox in humanity. On the one hand there is the unconditional value of humanity, like charity; on the other, the pragmatic imperative of historical, legal or political conditions that demand the opposite. These two remain irreducible.

1.2 On Medical Neutrality and Humanity

In order to understand the moral value humanity underlying the principle of medical neutrality, it is useful to start with Henri Dunant's classic humanitarian narrative, *A Memory of Solferino* (Dunant 1982). Witnessing the Battle of Solferino encouraged Dunant, who came from a devout and charitable Calvinist family, to find a way of preventing or at least attenuating suffering in future wars. Dunant was on a business trip in northern Italy in 1859 when the Battle of Solferino was raging nearby between the Austrian army and French and allied forces. More than 200,000 soldiers fought in that battle, resulting in more than 6000 deaths, over 20,000 wounded and 5000 captured or missing. When the towns and villages in the vicinity were filled with casualties and the army medical services proved inadequate, Dunant strove to organize care for the wounded and alleviate their suffering. Dunant thought highly of the French and Austrian soldiers, stressing their character, courage and humanity. He described the Croats and Hungarians, however, as savages and barbarians.

Inspired by his understanding of Christianity, Dunant has a moral sense of the importance of human life. Humanity appears to be a key ideal and value. He does not seem to be acting on a categorical imperative or absolute moral rule with no exceptions. Instead, his actions seem to be derived from a sincere concern for the lives of those soldiers. This concern is so strong that he even overcomes his own prejudices: he helps the 'barbarians' as well.

Dunant's book has two proposals: first, set up volunteer groups during peacetime to take care of casualties in wartime; second, convince countries to allow and support first-aid volunteers on the battlefield. He recommended setting up national relief societies and asked governments to protect and support them. In addition, he expressed the need for belligerent parties to declare field hospitals neutral in wartime, with similar protection extended to army medical staff, voluntary helpers and the actual wounded, and finally for governments to decide on a common distinctive sign indicating the persons and objects to be protected. Consequently, the International Red Cross Movement and the substantial set of universally recognized rules were developed, which now make up the Geneva Conventions. The original convention constituted the first attempt in modern times to codify the values Dunant described into an organizational ethos. Charity and universality, together with independence and impartiality were the essential and distinctive features of the Red Cross. These fundamental principles were first declared in 1920. The Red Cross Movement has always claimed that it does not raise its working principles, such as that of neutrality, to the status of absolute values: "The first principle – humanity – is considered the greatest principle, the motivating force and ideal, all other principles represent the means of achieving this aim" (Pictet 1956:12).

As such, the principle of neutrality aims to underpin the moral value humanity: Humanitarian aid should not be used for political purposes or to meet political agendas; it should be delivered without taking sides or engaging in controversies. Neutrality is not a moral value, but simply a form of outward behavior which demands self-imposed restraint; it means refusing to express an opinion concerning the qualities of the people or theories in question (Pictet 1956). Humanitarian or medical neutrality, therefore, does not equate to indifference or unprincipled relief workers; on the contrary, the underlying value is humanity, and neutrality is only the operational means of achieving this ideal in an essentially hostile environment (Slim 1997).

Moral neutrality in fact implies indifference; in the parable of the Good Samaritan, the (morally) neutral figure is the Levite who passes by, indifferent to the dramatic scene. Moral neutrality implies that the distinction between right and wrong loses its practical significance. The difficulty is that it seems impossible to separate operational and moral neutrality completely. Clearly, however, we cannot remain silent when facing the most heinous crimes. Confronted with such crimes, can human beings remain operationally neutral? If neutrality became an absolute, a "dogma" (Beigbeder 1991: 147), the ability to choose not to be morally neutral would be ruled out.

1.3 Unconditional Humanity

In this humanitarian narrative, at least in Dunant and Pictet's works, humanity appears to be unconditional. The principle of humanity stands on its own in the Red Cross doctrine and all other principles flow from it; as Pictet writes, "it is … its ideal, the reason for its existence and its object" (Pictet 1956: 14). Humanity is a synonym for charity: loving one's neighbor. It is a true altruistic form of love, the person who gives it is not considering their own happiness. Pictet explains that the ancient Greeks translated love as both eros and *agape*. Eros is the desire to appropriate something for oneself, whereas agape is altruistic, true disinterested love. According to Pictet, humanity should be seen as agape. He describes it as a feeling which "demands a certain amount of self-control; it may result from an effort [we are] required to make; its object may even be the enemy or a criminal" (Pictet 1956: 16).

1.4 Conditional Humanity

Several historical, legal and political events have confronted humanitarians with some of the most heinous crimes challenging the narrative of humanity as unconditional, true altruistic love or charity. For some humanitarians, these events also demanded the opposite, taking sides, or in other words: conditional humanity.

Can we argue that keeping silent, despite knowing about Nazi Germany's mass murder of Jews and others, made it possible to assist prisoners of war by sending them Red Cross food parcels and medicine, inspect conditions in the camps, trace missing soldiers, transmit messages between prisoners of war and their families and monitor compliance with the articles of the Geneva Conventions? As before, the International Committee of the Red Cross (ICRC) chose to uphold its neutrality in order to retain that access. After World War II, the ICRC faced severe criticism for this moral stance (Beigbeder 1991). We can ask however, what humanity demands in this situation, what events demand taking sides in order to uphold humanity?

During the Biafran War, the Nigerian military set up a food blockade around the nation's self-proclaimed and newly independent south-eastern region. Starvation was used as a weapon of war. Bernard Kouchner, a volunteer doctor within the ICRC, was frustrated by the delays in receiving permission to assist the starving Biafrans and by the ban on speaking out imposed by the ICRC. Having had members of his family murdered during the Nazi regime, Kouchner announced that he was not prepared to condone a second complicity in silence. He established *Médecins Sans Frontières* and opted for a more politically activist humanitarianism, openly denouncing oppression and injustice (Schweizer 2004). This 'new humanitarianism' (Hoffman and Weiss 2006) was based on the view that the old humanitarianism, which demanded operational neutrality in order to guarantee humanity as charity, posed a moral challenge when facing a situation where crimes against humanity were obviously being committed with intent to destroy, wholly or in part, a national, ethnic, racial or religious group. Concern for the victims, and hence for the ideal or value of justice, was essential for these new humanitarians. To keep silent and not to judge, in order to uphold humanity as charity—maintaining that even the worst criminal has a right to charity—was regarded as naive and unjust. The ICRC continued to argue that the price of denouncing injustice is usually expulsion from the scene of suffering, which in turn deprives victims of potential assistance and protection, and thus is likely to cause even greater suffering (Schweizer 2004).

Between 1994 and 1996, in the eastern part of the Democratic Republic of Congo (then Zaire), a 'refugee-warriors' community was established in refugee camps (Terry 2017). Humanitarians were uncomfortable about upholding unconditional humanity: they felt deep discomfort in assisting people who had murdered others, particularly since few of them expressed remorse. Additionally, the genocide against the Tutsis continued in the camps (Terry 2017). The President of Rwanda, Paul Kagame, argued that the refugee camps had been used as a base from where the former extremist (Hutu) government, army and *Interahamwe* militias launched raids on Rwanda (as well as in Zaire) to continue the killing that had started in April 1994. He declared the refugee camps a military zone. In 1996, across hundreds of sites in Zaire, Hutu refugees—men, women and children—were rounded up in camps, in forests or on roads, and either taken away, shot or slaughtered with machetes. While the Biafra context raised questions of whether aid could also prolong wars, in the aftermath of the Rwandan genocide it became clear that good intentions and charity or agape are simply not enough. Humanitarian aid can be manipulated as a powerful contributor to the destruction of a society (Cahill 2003).

2 Discussion

A similar paradox seems to be at play in the value of humanity as in forgiveness. There is the unconditional value of humanity as charity or agape; and at the same time, the pragmatic imperative that demands the opposite (i.e. taking sides). This paradox might even be more difficult for military medical personnel to deal with, especially in operational practice, where they often find themselves in a complex web of dependency (implying loyalty to certain groups) and power relationships. Accordingly, they have to deal with many different and potentially divergent moral responsibilities, often in isolated settings. As medical personnel are part of the military unit to which they are assigned, the military values underlining the importance of the mission, strategic thinking, and loyalty to the group may become increasingly important (Van Baarle et al. 2022). Unconditional humanity can be very hard to explain in an operational setting when the stakes are high, as decisions would be deduced from incontestable ethical precepts or principles. True humanity, like forgiveness, has to be related to a moment of unconditionality and requires respect for both extremes. The opposing extremes contradict one another but are at the same time fundamentally interwoven. It therefore presents immense difficulties for military health care personnel.

This paradox presents is unlikely to disappear, but this must not be an excuse for inaction. It calls for deliberation and understanding the almost unbearable demands implicit in the moral value of humanity that we have inherited from humanitarian traditions. We need to realize what it means when we strive for humanity, perhaps above all understand that humanity is a leap of faith that we must risk, but can never ultimately justify. It requires deliberation within and between the military and military health care organizations as well as ethics support for military health care professionals who are aware of their most fundamental moral values, so that they can make their position clear and are willing to stand up for the moral value of humanity.

3 Conclusion

This chapter focussed on the moral value humanity underlying the principle of medical neutrality (also referred to as political neutrality). Should it be considered an obligation, as being unconditional and in that sense, an absolute moral value? It seems that both the absolute and the relative are at play here. The contradictory logic in humanity appears similar to the paradox Jacques Derrida identified within the concept of forgiveness. There is the unconditional value of humanity, like charity, to provide medical care to those in need. At the same time, there is the pragmatic imperative of historical, legal or political conditions which demand the opposite (taking sides). These two remain irreducible: medical action thus has to be related to a moment of unconditionality to avoid being reduced to a strategic, political or economic calculation. On the other hand, such unconditionality can hardly be permitted in operational practice, as decisions would be deduced from incontestable ethical precepts or principles. It requires respect for both extremes. Military health care personnel are thus continually confronted with immensely difficult decisions: whether to judge, or condemn, who to deem innocent and where to show humanity and treat the sick and wounded.

References

Beigbeder, Y. 1991. *The role and status of international humanitarian volunteers and organizations: The right and duty to humanitarian assistance*. Vol. 12. Leiden: Martinus Nijhoff Publishers.

Benton, A., and S.E. Atshan. 2016. "Even war has rules": On medical neutrality and legitimate non-violence. *Culture, Medicine, and Psychiatry* 40: 151–158.

Cahill, K. 2003. *Traditions, values, and humanitarian action*. New York: Fordham University Press and The Center for International Health and Cooperation.

Derrida, J. 2003. *On cosmopolitanism and forgiveness*. London: Routledge.

Dunant, H. (1982). *A memory of Solferino*. Geneva: ICRC. Available at http://www.icrc.org/Web/eng/siteeng0.nsf/html/p036.

Gross, M.L. 2007. From medical neutrality to medical immunity. *AMA Journal of Ethics* 9 (10): 718–721.

Hoffman, P.J., and T.G. Weiss. 2006. *Sword & salve: Confronting new wars and humanitarian crises*. Rowman & Littlefield Publishers.

Jankélévitch, V. 1965. L'imprescriptible. *La Revue Administrative* 18 (103): 37–42.

List, J.M. 2008. Medical neutrality and political activism: physicians' roles in conflict situations. In *Physicians at war: The dual-loyalties challenge (237–253)*. Dordrecht: Springer.

Pictet, J. 1956. *Red cross principles*. Geneva: ICRC.

Slim, H. 1997. Relief agencies and moral standing in war: Principles of humanity, neutrality, impartiality and solidarity. *Development in Practice* 7 (4): 342–352.

Rubenstein, L.S., and M.D. Bittle. 2010. Responsibility for protection of medical workers and facilities in armed conflict. *The Lancet* 375 (9711): 329–340.

Schweizer, B. 2004. Moral dilemmas for humanitarianism in the era of "humanitarian" military interventions. *International Review of the Red Cross* 86 (855): 547–564.

Terry, F. 2017. *Condemned to repeat? The paradox of humanitarian action.* New York: Cornell University Press.
Van Baarle, E.M., C. Damsté, S.A. de Bruijn, and G.C. Bakx. 2022. Moral issues in soldier enhancement: Military physicians' perspectives. *Journal of Military Ethics* 21 (3–4): 198–209.
Van Tongeren, P. 2008. Impossible forgiveness. Ethical. *Perspectives* 15 (3): 369–379.
Wortel, E. 2009. Humanitarians and their moral stance in war: The underlying values. *International Review of the Red Cross* 91 (876): 779–802.

Chapter 5
Medical Neutrality as Impartiality—Implications for Prioritizing Medical Care in Armed Conflict

Erik Gustavsson

Abstract This paper focuses on medical neutrality as impartiality as a ground for rationing medical care in the context of armed conflict. Impartiality is often taken to entail that medical professionals should not discriminate between patients on "morally irrelevant grounds". Common examples of such grounds are age, gender, nationality, and political affiliation. This, in turn, is an idea akin to a principle of formal justice. Accordingly, medical neutrality as impartiality says something about how medical care should be distributed in armed conflict, or more specifically, it specifies grounds on which such care should *not* be distributed. However, it is unclear what this implies for rationing more specifically. The aim of this paper is to explore the concept of medical neutrality as impartiality, more specifically, to spell out the implications of medical neutrality as impartiality for the distribution of medical care in armed conflict. The paper argues that medical neutrality as impartiality must be understood as a mid-level ethical principle, rather than a high-level principle, to be applicable in practice. However, when ideas about formal justice are taken from high-level ethical theory and are applied on a mid-level, several questions arise. The paper then discusses two such questions.

1 Introduction[1]

The notion of medical neutrality in the context of armed conflict can be understood in various ways. For present purposes it is enough to distinguish between medical neutrality as either *immunity* or *impartiality* (Gross 2006). While medical neutrality

[1] This study was funded by Swedish Civil Contingencies Agency.

E. Gustavsson (✉)
Division of Philosophy and Applied Ethics, Department of Culture and Society, Linköping University, Linköping, Sweden
e-mail: erik.gustavsson@liu.se

D. Messelken, A. E. Barbar (eds.), *Challenging Medical Neutrality*, Military and Humanitarian Health Ethics, https://doi.org/10.1007/978-3-031-69398-4_5

as immunity refers to the claim that medical professionals should be given protection in armed conflict (or that they should enjoy some right to non-interference), medical neutrality as impartiality refers to the claim that medical professionals should not discriminate between patients on seemingly irrelevant grounds such as age, gender, nationality, and political affiliation. The focus for this paper is medical neutrality as impartiality, an idea saying that "morally irrelevant grounds" should not be taken into account when prioritizing medical care in armed conflict. This is an idea akin to a principle of formal justice. Traditionally, formal justice is often understood as a minimal formal requirement for theories about distributive justice and says, roughly, that like cases should be treated alike (Beauchamp and Childress 2018).

A straightforward interpretation of this idea in the context of armed conflict would be that treatment should be distributed based on medical need, irrespective of whether patients are compatriots, host-nation soldiers, civilians, or detainees. That is, while patient need is a morally relevant ground for the distribution of medical care, the fact that, for example, some patients are compatriots rather than civilians is not. This interpretation may seem straightforward. However, it is less clear what this idea implies for rationing medical resources in armed conflict more specifically. The aim of this paper is to explore the concept of medical neutrality as impartiality, more specifically, to spell out the implications of medical neutrality as impartiality for the distribution of medical care in armed conflict.

I will argue that in order to understand the implications of medical neutrality as impartiality it is important to interpret the idea as a mid-level ethical principle rather than a high-level principle. However, if understood accordingly, it raises questions with great importance for how resources are distributed that need to be discussed.

The paper proceeds as follows. In Sect. 2, I shall give some preliminaries. In Sect. 3, I shall spell out the difference between the principle of formal justice at a high-level and the mid-level. In Sects. 4 and 5, I discuss two questions that arise when the principle of formal justice moves from high to mid-level. In Sect. 6, I sum up and conclude.

2 Preliminaries

2.1 Medical Neutrality as Impartiality

I interpret medical neutrality as impartiality as an idea that says that "morally irrelevant grounds" should not be taken into account when prioritizing medical care in armed conflict. This is an idea akin to a principle of formal justice saying that like cases should be treated alike. In the words of Gross (2017): "Preferential care for compatriot warfighters stands in sharp contrast to the conventional principles of medical ethics that prohibit discrimination based on anything but medical need" (pp. 40–41). That is, rationing according to criteria other than medical need would be to violate the principle of medical neutrality as impartiality. For the sake of the argument, I will assume that the only morally relevant feature for rationing is medical need.[2]

Although distributing medical care according to need has been accorded a particular prominence in medicine, it is not always clearly stated how this idea should be understood more specifically or how it should provide guidance for rationing medical care.

Traditionally, in the literature on health care rationing, there are primarily two ways to understand distribution according to need. According to one interpretation, resource allocation should be based on how badly off a patient is. According to another, allocation should be based on a patient's capacity to benefit from treatment (Cookson and Dolan 2000; see also Hope et al. 2010). However, as I, and others, have argued more recently (Gustavsson 2018; Juth 2015), a more plausible interpretation of allocation according to need would be that such a principle is constituted by both how badly off a person is and a person's capacity to benefit from treatment. Principles of need will then ascribe weight to these two components to different degrees and for different moral reasons depending on how one interprets them more specifically.

As with the *relevant* grounds for rationing, the grounds that should be considered *irrelevant,* are open to discussion. Common examples of plausible candidates are age, gender, and nationality. Although these criteria may be morally irrelevant to rationing, they are sometimes used as grounds for rationing in practice.

For example, during the wars in Iraq and Afghanistan, American and British military medical facilities were guided by the medical rules of eligibility.[3] One of these principles dictates absolute priority to Coalition soldiers over civilians and non-Coalition soldiers: "...local nationals facing an imminent threat to "life, limb and eyesight" only qualify for treatment if...bed space is available at a combat support hospital that is not needed to treat Coalition forces..." (Gross 2017, p. 40).

This is a clear case of when the principle of neutrality as impartiality is not respected. Gross (2017) discusses various rationales (primarily military necessity, medical necessity, and associative obligations) for these rules in terms of how they can be justified. However, the aim for this paper is different from Gross' in that it does not focus on justifications for violating medical neutrality as impartiality. Rather, it is concerned with medical neutrality as impartiality and its implications. That is, before we discuss reasons for violating medical neutrality as impartiality, we need to know what it implies.

---

[2] This is how the conventional principles of medical ethics are usually understood. However, I believe, although I shall not argue for it here, that it is true for any system in which scarce resources are distributed that the cost-effectiveness of various treatments also should be taken as a relevant criterion for rationing.

[3] For a comprehensive discussion about the medical rules of eligibility see Gross (2017).

2.2 Further Preliminary Remarks

Before moving on to the discussion about formal justice I shall make a number of further preliminary remarks.

First, this paper is about prioritization. To prioritize is to give preference to one thing rather than another. In this context, this means giving preference to one patient's condition (combined with a treatment intended to meet that condition) over another patient's condition (combined with a treatment intended to meet that condition). The scarcity of resources implies that these kinds of priorities are always done, implicit or explicit. As a consequence, potentially beneficial treatment will be withheld from some patients. The withholding may take different forms. For example, postponing treatment, provide treatment that demands less resources, satisfying a need only to a certain degree (when it could have been optimally satisfied) or, simply, not treating a patient.

Second, as the subsequent discussion unfolds, I shall employ two kinds of examples. The difference being about the context. On the one hand, I will provide examples from a health care context. More specifically, a publicly funded health care system such as the one in Sweden, Norway, the Netherlands, UK etc. On the other hand, I will provide examples from the context of armed conflict.

Third, throughout this paper, I will often refer to resources being scarce. However, to refer to "resources" is not necessarily to refer to monetary resources but may also include medical professionals, competencies, time etc. That is, when referring to resources in the following, I refer to resources in a broad sense.

3 The Principle of Formal Justice—From High to Mid-Level

3.1 Formal Justice as a High-Level Principle

Traditionally, a distinction is drawn between formal and substantive principles of justice (Beauchamp and Childress 2018).[4] In the context of distributive justice, substantive theories about justice involve ideas about how goods ought to be distributed, whereas the principle of formal justice merely says that like cases should be treated alike; and are often thought of as a minimal formal requirement for theories about distributive justice. Formal justice is, in the words of Carr (1981) "…substantively empty and thus formally neutral between subjects" (p. 212).

[4] Substantive and formal principles can be further distinguished from procedural theories which are concerned with the process according to which decision-makers arrive at a certain decision regarding distribution. I shall leave such procedural accounts of justice aside in the following.

Medical neutrality as impartiality is an idea akin to a principle of formal justice.[5] A common way to formulate formal justice is the following: "Treat equal (like) cases equally (alike), and unequal (different) cases unequally (differently)" (Carr 1981, p. 211. See also Beauchamp and Childress 2018 for a similar formulation).

Accordingly, in its most abstract form, the principle of formal justice says that like cases should be treated alike. That is, at this level of abstraction, it does not provide any answer to the question about which respects individuals should be treated alike or what determines relevant (and irrelevant) differences (Beauchamp and Childress 2018). It merely says that morally irrelevant features should not be given any importance. This is why it is often described as a formal rather than a substantive principle of justice.

Traditionally, the principle of formal justice has been thought of as a logically independent consideration of justice (Rawls 1971, p. 58). However, Carr (1981) argues that while these considerations are indeed crucial for "doing justice" they need not be incorporated into a principle of its own (as is the case with the principle of formal justice). This is so because they are the logical implication of any substantive principle of distributive justice (Carr 1981, p. 224).

Consider next what is stated in World Medical Association's (WMA) handbook from 2022: "In emergencies, physicians are required to render immediate attention to the best of their ability. Whether civilian or combatant, the sick and wounded must receive promptly the care they need. No distinction shall be made between patients except those based upon clinical need." (WMA p. 136).[6]

So, according to WMA, health care resources should be distributed according to need, and nothing else. Consider next the claim made by Carr that "…it is quite unnecessary to think that we need a formal principle of justice to tell us that we ought to obey these material principles" (p. 222). If one applies Carr's reasoning in the context of the WMA handbook, it follows that considerations that are not based on peoples need-based claims should not be considered. This argument may be plausible when directed towards theories about justice on a high-level. However, when considerations about formal justice are referred to on a mid-level a number of questions arise. I shall spell out how these questions arise in the next section.

3.2 *Formal Justice as a Mid-Level Principle*

The focus for this paper is medical neutrality as impartiality in the context of armed conflict. That is not the same thing as formal justice on an abstract level. This difference can be understood in terms of high-level versus mid-level principles. High-level principles aim to provide general answers regarding what should be done and

[5] For an elaborated discussion about the relation between considerations about impartiality and considerations about formal justice see Carr (1981) p. 223.

[6] Similar wording can be found on p. 222.

why in all situations. To formulate mid-level principles rests on the assumption that it is useful and reasonable that decisions are guided by principles that are applicable in specific areas, in this case allocation of medical care in armed conflict. However, to characterize such mid-level principles is not to copy high-level theories and paste them into just any context. High-level theories are rather part of one's toolbox in that project. However, when the principle of formal justice is moved from high-level to mid-level, a number of questions arise. Before I move on to discuss these questions, consider two examples where ideas akin to formal justice have been formulated on a mid-level. These examples suggest that when formal justice is referred to in applied ethics, it is usually not characterized on such level of abstractness.

3.2.1 Example I—The Swedish Ethical Platform for Priority Setting

In Sweden, decisions about health care priority setting should be guided by an ethical platform (imbedded in Swedish legislation), consisting of three lexically ordered ethical principles:

(i) *The principle of human dignity.* All people have equal value and equal rights irrespective of their personal characteristics and function in society. Personal characteristics such as chronological age, gender, ethnicity, previous lifestyle, and societal function should *not* be considered in decisions about priority setting.
(ii) *The principle of need and solidarity.* Resources ought to be directed to patients or activities where needs are greatest. This principle also prescribes a striving towards levelling out differences in the population regarding opportunity and outcome with regard to health.
(iii) *The cost-effectiveness principle.* Decision-makers should pursue a reasonable relation between costs and effects when deciding how to allocate resources between different activities, measured in terms of improved health and increased quality of life.

Accordingly, when formal justice is referred to on the mid-level, it is usually not characterized with a high degree of abstractness. For example, when formal justice appears in the shape of human dignity in the Swedish platform it is far from being "substantively empty". On the contrary, it provides quite clear guidance for priorities in the sense that it specifies grounds on which priorities should *not* be set. The principle of human dignity clearly states the irrelevance of, for example, patient's previous lifestyle and their chronological age. Therefore, it seems incorrect to describe such a principle as "formal".

3.2.2 Example II—World Medical Association (WMA)

Interpreting medical neutrality in terms of formal justice in the context of armed conflict also specifies such conditions, for instance, the irrelevance of whether individuals are combatants or civilians. Consider next the way in which medical neutrality is expressed in the WMA handbook. WMA formulates medical neutrality as impartiality as follows:

"The medical duty to treat people with humanity and respect applies to all patients. The physician must always give the necessary care impartially and without discrimination on the basis of age, disease or disability, creed, ethnic origin, gender, nationality, political affiliation, race, sexual orientation, or social standing or any other similar criterion" (WMA 2022, p. 135).

Accordingly, impartiality in the context of priority setting in armed conflict, as in the context of priority setting in health care, clearly states the factors that should be considered irrelevant. That is, the principle of medical neutrality as impartiality also says something about how priorities should *not* be set in the context of war. Consequently, medical neutrality as impartiality has implications for how medical care should be distributed in armed conflict.

The two examples above suggest that when ideas about formal justice are taken from a high-level to a mid-level, they are no longer formal in the sense intended on the high-level. On the contrary, they are, indeed, substantive ideas about what should not be taken into account when it comes to the distribution of medical care.

The discussion above suggests the following: when ideas akin to formal justice appear on a mid-level a number of questions arise, questions to which different answers may have very different implications for how resources are distributed. I shall focus on two such questions in the following. First, that formal justice appears on a mid-level suggests that the principle no longer has the absolute status that it may have on a high level. This, in turn, suggests that formal justice should, just like other mid-level principles, be weighed against other principles. But what weight then, should be ascribed to formal justice as compared to other principles? Second, since formal justice on the mid-level is not just a formal "empty" principle, it (at least in the cases discussed above) comes with a number of features that should be considered irrelevant for prioritizing medical care. However, the features that appear on this list (intended to denote *morally irrelevant* features) may correlate with other, according to other principles for distribution of resources, *morally relevant* considerations. How should decision-makers relate to cases where *morally irrelevant* features happen to correlate with *morally relevant* features? In the following sections, I shall discuss these two questions.

4 The Weight of Medical Neutrality as Impartiality

To illustrate the way in which the question about weight arises, consider the following two examples. First, consider an example from a health care context borrowed from Sandman and Gustavsson (2016; see also Sandman 2023).

Example I. Suppose there are different diagnostic measures for detecting myocardial infarction in women compared to doing so in men. Suppose also that it is associated with a higher cost to detect myocardial infarction in women than in men. In such cases, Sandman and Gustavsson (2016) argue that the principle of formal justice implies that decision-makers should be prepared to accept a higher cost for the diagnostic measure for women than for men.

Consider next a similar case from the context of armed conflict.

Example II. It may be associated with a higher cost to treat wounded enemy soldiers compared to treating wounded compatriots. Suppose, for instance, that the enemy soldier poses a threat to health care professionals and needs surveillance. In such cases, medical neutrality as impartiality may imply that decision-makers should be prepared to accept a higher cost for treating the enemy soldier as compared to the compatriot.

These two examples reveal an important feature about formal justice on a mid-level. Principles akin to formal justice on a mid-level are often interpreted as absolute in the sense that it is *never justified* to violate them. The idea seems to be that these considerations should work as deontological constraints on other considerations. Consequently, *no weight* should be given to which party to the conflict patients belong to. However, the implications of such constraints may have unreasonable implications. While it may be reasonable to accept a higher cost to treat enemy soldiers (compared to compatriots), it seems unreasonable to accept any cost. Accordingly, as soon as formal justice is moved from a high-level to a mid-level, the grounds considered to be irrelevant to priority setting are specified. This means, in turn, that there will be an ethical cost to distribute resources in accordance with mid-level formal justice. This line of reasoning reveals something important about medical neutrality as impartiality, namely that it cannot reasonably be interpreted as an absolute principle. Accordingly, medical neutrality as impartiality should be interpreted as a mid-level principle for distributing medical care in armed conflict. Accordingly, a patient's claims based on medical neutrality as impartiality may have to be weighed against that patient's need-based claims (and perhaps other relevant considerations).

5 Correlation with Other Criteria

The principle of human dignity (in the Swedish ethical platform) and the considerations of medical neutrality made in the WMA-handbook constitute two examples of when considerations about formal justice have been characterized on a mid-level. Furthermore, when considerations about formal justice are spelled out on a mid-level there are a number of typical examples of criteria combined with the principle of formal justice, criteria that should not be considered. However, different criteria added to the list of *morally irrelevant* features may often correlate to different degree with *morally relevant* aspects, even if they themselves are not perceived as such. This section focuses on two such criteria: age and nationality.

To illustrate the way in which the question about correlation arises, consider the following two examples.

Example I—Age. A view according to which expected life years should be maximized would often give preference to the young rather than the elderly. However, this is not *because* of their age but because the life expectancy of the young normally is longer than for the elderly. That is, the justification is that gained life years are maximized, it just so happens that expected life years often correlate with young age.

According to WMA, age should not be considered. Does that rule out other considerations that happen to correlate with age? Suppose that this question was answered in the affirmative. Consider then the aforementioned assumed morally relevant feature for rationing: medical need, which, in turn, is matter of how badly off one is as well as how much one can benefit from treatment. In rationing, this would often take the form of expected benefits that accrue to a patient (or patient group). These benefits may be improvements in longevity or quality of life. Consider next the former rather than the latter. It goes without saying that a young patient often has more life years to gain than an elderly patient. That is, a treatment that targets a young person may provide a health improvement (in terms of longevity) of, say, 70–80 years, whereas this is not the case for treatments targeting elderly patients. Accordingly, to exclude considerations about improvements in terms of longevity from considerations about rationing seems to have implications that are difficult to accept. This suggests that mere correlation with morally irrelevant criteria cannot reasonably be a reason to exclude such considerations.

Example II—Nationality. The mere fact that the person in need is a soldier from an affluent country, rather than an indigenous civilian in some poor rural area, should not, in itself, be a reason to withhold care for that civilian. However, there are ways in which the soldier may be different from the civilian from a priority setting perspective.

For example, it will often be more uncertain whether the civilian will have access to the necessary follow-up care. In some cases, follow-up care may even be necessary for any benefit to accrue to the patient. Therefore, (1) the risks may be too great in relation to the potential health improvement, and (2) the improvement may be too small in relation to the resources it requires.

However, from an ethical perspective it is still important to stress that the underlying justification for why the soldier should be prioritized is not *because* he is a soldier rather than a civilian but because the effect of whatever treatment the two individuals need—the benefit will be very different for the two. The difference in patient benefit depends on whether the patient has access to the necessary follow-up care. To consider the patient's availability of follow-up care relevant would not be unique for the soldier versus civilian case. In other circumstances health care professionals do ascribe importance to the possibilities for after care. For example, organ transplant and likelihood to adhere to treatment plan.

Accordingly, the mere fact that a patient is a soldier, rather than a civilian, should not, in itself, be a reason to withhold care for that civilian. However, given that the soldier has better prospects in terms of receiving the relevant follow-up care the treatment given to the soldier will, in fact, be more beneficial. Therefore, there is a morally relevant reason to prioritize the soldier rather than the civilian.[7] However, this is not because the soldier is a soldier, and the civilian is a civilian but because these attributes happen to correlate with prospects of receiving relevant follow-up care.

Accordingly, a reasonable interpretation of the claim that nationality should be considered a morally irrelevant feature would be that while decision-makers should not consider nationality "as such" a relevant criterion for the distribution of resources, it does not rule out morally relevant considerations that happen to correlate with the irrelevant considerations.

Note that this conclusion is compatible with the claim made above, namely that medical neutrality as impartiality must be weighed against other principles and that this may involve a cost in terms of other ethical considerations.

6 Conclusion

In this paper I have explored medical neutrality as impartiality and spelt out its implications for the distribution of medical care in armed conflict. I have argued that medical neutrality as impartiality should be understood as a mid-level principle for prioritizing medical resources in armed conflict. It is important to note that medical neutrality as impartiality is not merely formal on the mid-level but has real implications for how resources should be distributed as it specifies grounds according to which rationing should *not* be done. Since the irrelevant aspects do not fall from the sky and because they play an important role for rationing medical care it is important to discuss which these criteria should be. Consequently, medical neutrality as impartiality cannot reasonably be interpreted as an absolute principle. That is, it will in some cases have to be weighed against other principles. Furthermore, the paper discussed cases in which morally *irrelevant* features happen to correlate with morally *relevant* features and argued that medical neutrality as impartiality cannot reasonably be concerned with such considerations.

[7] However, note that these implications depend on that one accepts the background conditions. That is, it is still a wider ethical question, and there may be good reasons to believe, that it is unfair that the possibilities for follow-up care are unequally distributed in the first place.

References

Beauchamp, T.L., and J.F. Childress. 2018. *Principles of biomedical ethics*. Oxford: Oxford University Press.

Carr, C.L. 1981. The concept of formal justice. *Philosophical Studies* 39 (3): 211–226.

Cookson, R., and P. Dolan. 2000. Principles of justice in health care rationing. *Journal of Medical Ethics*. 26 (5): 323–329.

Gross, M.L. 2006. *Bioethics and armed conflict moral dilemmas of medicine and war*. Cambridge, Mass: MIT Press.

———. 2017. Saving life, limb, and eyesight: Assessing the medical rules of eligibility during armed conflict. *The American Journal of Bioethics* 17 (10): 40–52. https://doi.org/10.1080/15265161.2017.1365186.

Gustavsson, E. 2018. *Characterising needs in health care priority setting*. Doctoral dissertation. Linköping University, Linköping.

Hope, T., L.P. Østerdal, and A. Hasman. 2010. An inquiry into the principles of needs-based allocation of health care. *Bioethics* 24: 470–480.

Juth, N. 2015. Challenges for principles of need in health care. *Health Care Analysis* 23: 73–87.

Rawls, J. 1971. *A theory of justice*. Cambridge, MA: Harvard University Press.

Sandman, L., and E. Gustavsson. 2016. The (ir)relevance of group size in health care priority setting: A reply to Juth. *Health Care Analysis* 25 (1): 21–33.

Sandman, L. 2023. In defense of a 'thick' formal equality principle in healthcare resource distribution. *De Ethica* 7 (4): 50–67.

Chapter 6
Medical Neutrality

Paul Gilbert

Abstract Under international law military doctors are expected to be neutral in the sense of treating the wounded in accordance with their medical needs alone irrespective of which side they are on, and military staff enjoy immunity from attack in consequence. But it has been objected that their military role should sometimes lead them to act otherwise. The paper maintains that this misunderstands the doctor's role in the armed forces, but the objection does raise the question of why the forces employ doctors with their strict ethical code rather than medical technicians who would just obey orders rather than following purely medical priorities. The paper claims that this is because adherence to the principles of discrimination and proportionality would result in such staff usually being ordered to act in accordance with the same priorities as doctors proper. The immunity from attack of medical staff can be derived from the same principles. Possible exceptions and the applicability of the present system of neutrality to contemporary conflicts are discussed.

1 Introduction

I shall start with a story. An injured man is helped into a doctor's surgery. The doctor recognises him as one of her patients. But before she can begin treating him someone rushes in shouting, "There are many people down the road in a worse state than

First published: Gilbert, Paul. 2015. Medical Neutrality. In *Proceedings of the 4th ICMM Workshop on Military Medical Ethics*, ed. Daniel Messelken and David Winkler, 10–23. Bern: Bundespublikationen.

Paul Gilbert passed away in 2022. His chapter, republished here, was a contribution to an earlier workshop, but has lost none of its topicality, and the reflections on medical neutrality complement this anthology perfectly.

P. Gilbert (Deceased) (✉)
University of Hull, Hull, UK

D. Messelken, A. E. Barbar (eds.), *Challenging Medical Neutrality*, Military and Humanitarian Health Ethics, https://doi.org/10.1007/978-3-031-69398-4_6

him"—pointing to the injured man—"and he's the one who caused this disaster. Come and help them. Please!" We can readily imagine that the doctor would respond to this appeal, considering that the apparently greater medical needs of the unknown people down the road take precedence over that of her patient.

We might think of several possible civilian contexts for this scenario, but I want instead to paint in a military background for the story. The doctor is a military physician at a field hospital. The injured man brought in is a soldier from her own side wounded in a street battle. The people in a worse state down the road are civilians from the enemy side injured when he fired a rocket propelled grenade at a house from which there had been sniper fire. Might these circumstances change the doctor's priorities regarding whom she should treat from those that are operative in a civilian context? Does she face a dilemma special to the military situation, bearing in mind that if promptly treated the soldier may be able quite soon to return to fighting?

It is a cardinal principle of international humanitarian law as set out in the 1949 Geneva Convention that wounded soldiers from either side are to be treated without distinction as to their affiliations, but only in accordance with medical priorities, and in the 1977 Additional Protocols this principle is extended to the treatment of injured civilians. In recognition of the fact that military medics must be neutral in the way that they select patients for treatment they are accorded immunity from attack—medical neutrality in the legal sense. They are regarded as non-combatants and enjoy protected status in the same way as if they were civilian doctors because they are expected to act like them. Legally, then, it seems clear the military doctor in our scenario ought to do whatever her counterpart in the civilian context would do. Yet, we may ask, is this ethically the case also?

2 The Role of a Military Doctor

In recent years there have been several suggestions to the effect that the situation of the military doctor is more complicated than that of the civilian one, in particular in view of the fact that the former occupies two roles, both a military and a medical role (see e.g. Allhoff 2008). This has led to the conclusion that an army medic should sometimes act otherwise than in accordance with purely medical priorities on account of the divergent ethical requirements of the military role. But the supposition that the requirements of the two roles may conflict is, I want to argue, false since it fails to allow that the armed forces, like other employers, employ doctors just as doctors, not in some other sort of role for which their knowledge and skill equips them. The role of doctor is self-standing, like that of a philosopher, say, in the sense that pursuing the calling is possible outside of an institution. When it is part of an institution's workings the institution treats the role as ready-made and plays no part in shaping what is involved in performing it. Connected to its self-standing status are two further features of the doctor's role. First, it is pervasive in that a doctor is expected to give medical assistance in an emergency even outside of his or her employment. Second, it is autonomous in that the doctor is solely responsible for the medical judgments on which he or she acts and should not be ordered to act otherwise by an employer for non-medical reasons.[1]

If this is right then it is evident that a military doctor will have to act on his or her medical judgment alone when acting as a doctor, and this will imply neutrality in the treatment of the wounded, for enemy wounded may demand more immediate treatment than the doctor's comrades. It is not for the doctor, except, as I shall later allow, in the most desperate circumstances, to make a moral judgment about what it is best to do all things considered. It is not for the doctor to take account, say, of the military consequences of the medical priorities decided upon, of the relative deserts of the wounded or of any other non-medical factor. One way to express this is to say that the doctor's role gives its occupant a reason not to act in situations where medical judgment is required otherwise than for medical reasons, that is to say that it excludes acting from non-medical considerations, and thus occupying the role constitutes having what Joseph Raz terms an exclusionary reason (Raz 1990). Medical neutrality in the sense of treating patients irrespective of whether they are friend or foe is simply a consequence of the military doctor acting just as a doctor and thus following medical priorities determined by the severity and so forth of the patients' conditions.

Yet, someone may reply, surely as a member of the armed forces the military doctor is committed to obeying orders, and these orders give him or her an exclusionary reason not to act from any other consideration. Does this military role not create the potential for a clash of exclusionary reasons, if, for example, a doctor is ordered to treat lightly injured comrades first and let treatable enemy fighters die? What we should say here, I suggest, for reasons that should become clearer later, is that such an order is an illegitimate one, as bad in its way as an obviously unlawful order to kill the enemy fighters. At this point, however, we can raise the question why the forces should employ some people who, if I am right, may sometimes have an exclusionary reason not to obey what they are ordered to do—people who in this way fall outside the ordinary chain of military command. For there may seem to have been an alternative to this, namely for the forces to employ medical staff to act, just as tank mechanics do, to patch up damaged parts of the fighting machine for further service now or later in the war. Why do the armed forces employ doctors, with their stringent ethical code which might lead them to resist orders, rather than mere medical technicians who possess the knowledge and skill of doctors but who are committed to obeying commands?

[1] For development of these points see Gilbert 2013.

3 Doctors or Schmoctors?

Following Arthur Isak Applbaum, who introduces this notion in a different context, we can call these imagined medical people “schmoctors” (Applbaum 1999: 48–51). Schmoctors differ from doctors in not being autonomous but in following out the orders of their employers, in this case the military. The schmoctors’ role is not self-standing either, since without employers to instruct them they have no function, nor, correspondingly, is it pervasive since they are not expected to use their skill and knowledge outside their employment. We may suppose, for the sake of argument, that schmoctors may be instructed to prioritise the treatment of lightly wounded troops to return them service as quickly as possible, and to give preference to caring for their own side’s casualties, contrary to the medical priorities that doctors would follow. Why, then, are doctors employed by the military rather than schmoctors?

The first and most obvious answer to the question why the military employs doctors rather than creating a cadre of schmoctors might seem to be that there are already many trained doctors around to be employed. But it may be idealistic to think that not enough of them would abandon adherence to their strictly medical ethic in order to become schmoctors and follow out military rather than medical priorities. Nor would this choice need to be regarded as unethical if the doctors who became schmoctors were taken to attach greater moral significance to a military role than to a conventional medical one—a supposition not implausible in some wars. Yet this is really not the sort of answer to our question that we are seeking, for what we want to know is why the military might think it better to employ doctors rather than schmoctors.

A second suggestion gives, then, a prudential reason, namely that doctors are employed rather than schmoctors because doctors are immune from attack in virtue of being neutral in the treatment they give to the wounded, whereas we can assume that schmoctors would not be immune any more than tank mechanics are, since they are similarly acting simply to contribute to the war effort. Immunity from attack would certainly allow lightly wounded troops to return to fighting without hindrance, but it is far from clear that the advantages of this would outweigh the disadvantages of doctors prioritising other more serious casualties, including the enemy’s. Furthermore, this prudential account presupposes that a system providing immunity from attack for military medics is already in place, while this is one of the things that itself needs to be explained. It is not obvious that the even-handedness of doctors by itself warrants their immunity. For, even if they are neutral, the effects of their behaviour may not be, since, while their wounded comrades may be returned to fighting or at least be enabled to participate at a later stage in the war, the enemy’s casualties are held as prisoners of war. There might, therefore, be military advantages in targeting medical facilities even if it led to the death of some of one’s captured comrades, particularly if one’s opponents’ medics were more effective in maintaining troops in the field than were one’s own. It is not clear that the system of medical neutrality can be justified in prudential terms, so the question remains why it is doctors, with their ethical norms which require neutrality, who are employed by the military rather than schmoctors, who lack these norms.

4 Military Ethics and Medical Ethics

The answer I shall go on to suggest is an ethical one. It is that doctors are employed because their medical priorities coincide, broadly speaking, with the priorities which military commanders should have if they are conforming to the traditional laws of war. I shall assume without argument that these laws have an ethical force, and they are, of course, to a large extent embodied in international law. My argument will be, then, that schmoctors do not need to be employed because the orders they ought to be given would usually coincide with the judgments about how to act that doctors would make anyway. How might we reach this perhaps startling conclusion?

The two traditional laws which regulate the conduct of armed conflict are the principle of discrimination, which forbids the targeting of non-combatants, and the principle of proportionality. The latter principle has two aspects: first, it forbids the infliction of harm which does not promise any military advantage, which is, in other words, not militarily necessary; second, it forbids action which is more harmful than is warranted by the military advantage expected from it, so that sometimes a specific advantage should be foregone because the means to achieve it are too harmful. While in codified international law the principle of proportionality is restricted to the way in which civilians are affected, we may regard it as having a wider scope ethically, taking in also disproportionate attacks on enemy troops.[2]

The principles of discrimination and proportionality can be regarded as giving more specific shape to two ideas underlying our thinking about the ethics of war. One is that the harm that war inevitably causes should be kept to a minimum, the other is that the combatants in war are fighting, at least in part, to protect non-combatants, particularly civilians. Thus the principle of discrimination reduces possible harm by limiting those who can be targeted, while the principle of proportionality seeks to minimise harm when attacks are made on legitimate targets. The principles of discrimination and proportionality both aim at the protection of non-combatants on either side, so that combatants who wish to protect their own civilians can agree to a regime which incorporates these rules as helping them to accomplish this, even if it sometimes restricts the scope of military activities in which they seek to protect their own civilians more directly. Adherence to the principles thus demonstrates the concern for civilians caught up in conflict that combatants are expected to possess themselves and should be able to assume in their opponents.

It is convenient to start with the protection of civilians in developing an argument for the employment of doctors with their specifically medical priorities, despite the fact that the 1977 Additional Protocols requiring them to be given medical treatment post-date the Geneva Convention provisions mandating impartial medical attention for wounded troops. For we can see that if combatants have a reason for following rules which protect civilians on both sides then they have an equal reason for having civilians injured in conflict treated, since this has the effect of protecting them from

[2] For discussion of proportionality see Gilbert 2014.

more of the harmful consequences of war. There is, though, a more specific reason why military forces should require their medical personnel to treat such civilians. It is that the principle of proportionality requires that the scale of civilian casualties should be minimised in military operations. But if these casualties are caused and yet their effects could be alleviated by medical intervention then it would seem that the principle of proportionality demands that this intervention is forthcoming when it is feasible. If the forces that cause civilian casualties have medical personnel available then it is their job to treat the injured and this will usually accord with what their professional ethic requires of them.

"But", it might be replied, "there may be a need to employ the medics in patching up troops for fighting and this might be a military priority." I have to reply that getting lightly wounded troops back into action as a priority will usually either not be necessary or will not secure a sufficient military advantage to warrant the death and suffering of the civilians affected by treating these troops first. This claim is, of course, partly an empirical one about what the requirements in war are likely to be. However, it is also an evaluative claim about the ethical weight to put upon avoiding harm to civilians. To the extent to which this is regarded as a supremely important value then the value set upon achieving particular military objective may diminish. Thus just as one's own forces may have to accept casualties in order to minimise collateral damage to civilians (Walzer 1977: 155–6) so also they may need to accept that their own medical treatment should sometimes take second place to the treatment of civilians whom their operations have injured. Similarly, immediate military operations may need to be rethought in the light of the need to treat such civilians when their treatment delays attention being given to the soldiers required to secure the objectives of these operations.

5 Treating Wounded Soldiers

So far I have tried to justify the military employment of doctors with their medical priorities on the grounds that these priorities may usually be expected to coincide with the priorities demanded by a serious application of the principle of proportionality concerning harm to civilians. But we can extend the argument to cover enemy soldiers who are rendered *hors de combat* by their wounds and who are thereby also classified as non-combatants. The point of targeting enemy troops is simply to put them out of action as soldiers, not to inflict upon them additional suffering, injury or death. To do so is to cause unnecessary harm. Similarly, to withhold medical treatment, which would prevent such harm, is itself to cause unnecessary harm. In general, then, an operation which results in enemy troop casualties that could have been treated but are not may be regarded as disproportionate unless there is a very strong countervailing need to get one's own lightly wounded soldiers back to the line quickly. If this is, as I claim, seldom the case if harm to non-combatants is given its due ethical weight, then again a doctor's medical priorities in treating the seriously wounded enemy troops first will usually coincide with the requirements of adherence to the principle of proportionality.

In the case of badly wounded soldiers we can see that combatants will generally have a double motive for enabling them to be given urgent medical treatment. First, they have the same motive as they have for enabling injured civilians to be treated, namely that this serves to protect them from some of the harms of war, since soldiers hors de combat are also non-combatants to be protected just as civilians are. Second, however, combatants have an interest not only in those they protect being given medical treatment but also in being treated themselves when they are wounded, irrespective of whether they find themselves within their own lines or captured by the enemy. Just as we can think of combatants as agreeing to rules of war that prevent them from being subjected to disproportionate harm (Gilbert 2005) so we can also suppose that they would regard being given treatment for their wounds if captured as a proper application of such principles. But falling in with these practices of war is not only self-interested, it also demonstrates a concern for the fate of fellow combatants whatever their affiliation—an attitude which contrasts with vengefulness or hatred against them as the enemy. It therefore expresses an ethical attitude as well as a prudential one, and it is such an attitude that is captured in the laws of war regarding the treatment of prisoners generally.

To return, then, to my question of why it is doctors whom the military employs and not schmoctors, who, we supposed for the sake of argument, might be instructed to prioritise returning troops to battle irrespective of what medical ethics prescribes. My answer is that normally these ought not to be the instructions that are given. For adhering to the laws of war will usually require that badly injured individuals are prioritised whatever their affiliation or their status as civilians or wounded soldiers. But this is what would accord with a doctor's medical priorities. So schmoctors are not required since doctors' priorities normally coincide with what schmoctors ought to be instructed to do anyway.

6 Killing and Allowing to Die

Before moving on, however, we must confront an objection that could be made to my account, which concerns what exactly the principles of discrimination and proportionality do demand. The former principle enjoins, for example, that a badly wounded enemy soldier should not be deliberately killed. This, the objector may observe, does not forbid him being allowed to die by withholding medical assistance. And so, it could be held, for other instances of discrimination and proportionality *pari passu*, that I have required too much by way of adherence to these principles.

This is not the place for a general discussion of the moral significance of the doing/allowing distinction, but in the kind of case we are considering we need to go back to the ideas about the conduct of war that underlie the discrimination and proportionality principles in order to see why the objection may seem casuistical in the pejorative sense of the word. For these ideas envisage a regime in which unnecessary harm is avoided and those not directly involved in the conflict are protected as much as possible from its harmful consequences. In the example above, therefore, allowing the soldier to die when medical treatment would have prevented it does cause unnecessary harm and leads to his death as surely as if he had been killed outright. In circumstances where treatment is available it is hard to escape the conclusion that, other things being equal, the intention is that he should die. It is this sort of intention, from whatever motives it springs, that the principles of discrimination and proportionality are designed to forestall. It is, however, precisely the sort of intention that a doctor, with his or her commitment to saving life, will eschew.

7 Immunity from Attack

I think we can see from the account I have offered why military medical personnel should enjoy medical neutrality in the legal sense of immunity from attack. First, we can see that if armed forces have a duty to mitigate the effects of war as much as possible then attacks upon medical staff who are doing this in their own particular way would hinder or prevent the discharge of this duty; freedom from attack is a necessary condition for medics to treat the wounded effectively. Second, any such attack would be likely to constitute an indirect attack upon non-combatants being treated. For, even if they are not being directly targeted, depriving them of medical aid by attacking medical facilities may have the same harmful effects and thus breach the principles of discrimination or of proportionality. It should be added that since those being treated may include the attackers' own comrades or compatriots there is a further prudential reason to avoid attack, though one that is operative only because the medics are neutral as between friend and foe in the treatment they provide.

Military medics count as non-combatants themselves and as such may not be targeted. And they do so count just because of the way they treat the wounded irrespective of their affiliations. But, it will be objected, military medics do not only treat seriously injured soldiers; they also treat lightly wounded combatants to return them to fighting, and this is a direct contribution to the war effort. Why does this not disqualify them from immunity as much as if they spent some of their time firing at the enemy, for this normally would? The reason, I think, is that there is no such sharp distinction between treating badly wounded and lightly wounded soldiers as there is between treating non-combatants of whatever sort and actually fighting. Perhaps armed forces would not employ medical staff only to treat the badly wounded even if the distinction could be made, though arguably it is incumbent upon them to employ staff who do treat the badly wounded if they are to conform to the principle of proportionality. Perhaps, however, armed forces do in fact employ doctors precisely because medics can "conserve the fighting strength", as the motto of the US medical corps has it. But given this unavoidable fact, it is still better to regard them as non-combatants than as combatants, in the way that tank mechanics are, because they do treat non-combatants and treat them irrespective of their allegiance in accordance with their medical needs.

However, I believe that the fact that military medics are accorded immunity from attack does itself have further consequences for what their priorities should be. Since they gain immunity because they treat non-combatants then it is these, I suggest, who have a *prima facie* morally prior claim to treatment, whether they be civilians injured in hostilities or badly wounded fighters. Though this priority both usually coincides with what the medics' professional ethic mandates and, I claim, can be derived from other, military considerations, the fact that immunity is offered as a *quid pro quo* for neutrality in medical treatment is a yet further reason for exercising such neutrality. This is not only a prudential reason but also an ethical one, since this sort of behaviour and the immunity it gives them are both expressions of the medics' humanitarian role in war.

8 Battlefield Triage

If my account is correct then it is not acceptable that reverse triage should be normal in wartime situations, with the salvage of lightly injured troops on one's own side being prioritised over the treatment of seriously hurt soldiers and civilians from both sides. This would put the military function of medical staff ahead of their more strictly medical and humanitarian one, which is to overlook the reasons for which such staff gain immunity from attack, namely that they mitigate the harms of war despite their contribution to its conduct. To the extent to which the latter comes to override the former the reasons for their immunity begin to fall away.

It could be objected that combatants might agree to having their own return to fighting prioritised over the treatment of more seriously wounded soldiers and civilians, so that reverse triage is quite acceptable. But it seems unlikely that they would agree to it as a general rule, not least because, if I am right, it is hard to square with the principles of discrimination and proportionality to which combatants are taken to have agreed. For reverse triage will usually cause more suffering to non-combatants than that to which they would otherwise be exposed, and so it is unacceptable unless, that is, the action into which salvaged troops are sent would prevent a greater degree of harm. But this, I claim, will surely be the exception.

Medical priorities do, of course, take account of patients' wishes as well as their wounds. A parent, for example, may want his less badly hurt child treated before himself. Similarly a badly wounded soldier may consent to his less hurt comrade being treated first so as to carry on the fight. Perhaps soldiers may even be presumed to have consented to these priorities in virtue of having accepted that it is their function to do all they can to secure victory and to bear the costs of this. Yet this does not imply that soldiers and still less civilians have consented to such priorities in the event that they are badly wounded and find themselves in the hands of the enemy. Presumably they would not do so, and then the general presumption against reverse triage stands.

9 Exceptions to Neutrality?

I would not wish to claim that a military doctor should follow the usual peacetime priorities in every conceivable wartime situation. We can envisage desperate circumstances where a ruthless enemy would break through the lines and deal death and destruction, including to civilians, unless some lightly wounded soldiers were returned to battle at the cost of some badly injured people dying as a result of lack of treatment. But if this is the choice a doctor makes it is not because he or she also has a military role. One can imagine a civilian doctor making the same judgment; and one can compare this situation with the circumstances in which a doctor is placed when confronting the effects of a natural disaster in peacetime, where he or she may prioritise keeping slightly hurt rescuers working to maximise the number of victims who are saved at the expense of letting some who are seriously hurt die. Notice that in the wartime case no appraisal of the value of attaining a military objective is involved. The calculations made are simply humanitarian ones like those in peacetime, consistent with the doctor's calling of saving lives and alleviating suffering.

It follows that it is not the business of the military doctor to make assessments of proportionality in which the lives of wounded enemy civilians, which could be saved if resources are not devoted to returning troops to battle, are measured against the importance of the military objective, which their return to fighting should secure. This would contravene the doctor's duty to act in accordance with strictly medical priorities—a duty he or she must follow except in the most desperate circumstances, though even here the humanitarian purposes that lie behind these priorities provide the doctor's guide.

One should, however, resist any suggestion that medical staff might legitimately lose their immunity from attack if they act in ways that put military priorities before medical ones. There is a distinction between those who break the rules of a role and those whose actions do not count as performances of the role at all, though there is often no sharp division here. But the way to treat those who break the rules is to apply sanctions to them if and when this becomes feasible, and this is to continue to regard them as role performers. Military medical personnel should be regarded for as long as possible as performing their medical role, even if defectively, so that only if they actually join in the fighting or they grossly violate the rights of wounded prisoners and civilians should they lose their protected status. Here sharp lines have been crossed where before no sharp line could be discerned, and in the absence of such a line it would be perilous to withdraw immunity from doctors who may be acting under duress or other intolerable pressures to depart from their professional norms. The whole system of medical neutrality would be under threat if combatants could judge whether military needs were being given preference over medical ones and attack medics who failed thus in their duty.

It would be a different matter if medical neutrality were a mere convention, so that if the rules of the convention are violated by one side then they may legitimately be broken by the other. The picture I have painted does not imply that all we have here is a mere convention, adopted for convenience to allow for the treatment of the wounded which would be difficult if medical staff came under attack. If this is all that the system of medical neutrality came to then a failure of reciprocation in adhering to the rules by one side might justify the other in breaking them, and so the motive for adherence would be largely a prudential one. Evidently, then, regarding neutrality as a mere convention would affect the attitude of medical staff towards it, for they would not need to feel any ethical obligation to conform to its rules. This is not, however, how military medics do view the demands of their role, rather seeing their actions as having a moral grounding in the performance of it as a humanitarian vocation—a vocation whose requirements in conflict can, I have argued, also be derived from just war principles.

10 Contemporary Conflicts

All this may seem wildly idealistic in an era when medical neutrality is frequently flouted in that both captured fighters and injured civilians are denied medical assistance, ambulances are targeted and so forth. More generally, the value of saving non-combatants from preventable death or injury relative to that of securing immediate military objectives is arguably often set too low, with indiscriminate or disproportionate attacks being commonplace. The context of many of these cruelties is often one of civil conflict and irregular or asymmetric warfare, involving groups espousing different identities or values. Yet the fact that it is in just these circumstances that violations occur seems to me to support my account of the ethical grounding of medical neutrality, since in these circumstances we cannot rely upon combatants having the ethical attitudes to which I have appealed in justifying the practice.

Whereas in most conventional wars combatants do see themselves as acting to protect their own civilians and can agree to rules that offer a degree of protection to civilians generally in the furtherance of this aim, in many contemporary conflicts, by contrast, combatants are not prepared to regard both their own civilians and those from an opposing identity or value-based group as entitled to the same protection. Unlike war on Rousseau's account as "something that occurs not between man and man, but between States" (Rousseau 1762: bk.1, ch.4), these contemporary conflicts are precisely 'between man and man', with the principles of discrimination and proportionality widely ignored because members of the opposing group inspire hatred, resentment, anger or contempt. Combatants in these conflicts, then, may lack the ethical motivation either to facilitate the treatment of their opponent's wounded or to refrain from targeting those who are treating their own casualties. Faced with combatants of this character there may well be a temptation to reply to their violations in kind. But, I have argued, there are strong ethical reasons to uphold the system of medical neutrality as we have it and not to let it become distorted or to die out through being abused.

References

Allhoff, F., ed. 2008. *Physicians at war: The dual-loyalties challenge*. Dordrecht: Springer.
Applbaum, A.I. 1999. *Ethics for adversaries: The morality of roles in public and political life*. Princeton: Princeton University Press.
Gilbert, P. 2005. Proportionality in the conduct of war. *Journal of Military Ethics* 4: 100–107.
———. 2013. Medical neutrality and the dilemmas of war. In *Military medical ethics for the 21st century*, ed. M.L. Gross and D. Carrick, 84–95. Farnham: Ashgate.
———. 2014. Understanding proportionality in contemporary armed conflict. In *The Ashgate research companion to military ethics*, ed. J.T. Turner and E. Patterson. London: Routledge.
Raz, J. 1990. *Practical reason and norms*. Princeton: Princeton University Press.
Rousseau, J-J. 1762. *The social contract* (Many editions).
Walzer, M. 1977. *Just and unjust wars*. New York: Basic Books.

Part III
Medical Neutrality in Practice

Chapter 7
Medical Neutrality or Medical Humanity in War?

Martin C. M. Bricknell

Abstract The phrase medical neutrality is founded upon the Humanitarian Principles of humanity, neutrality, impartiality, and operational independence that apply to United Nations agencies and humanitarian non-government organisations during disasters and complex humanitarian emergencies. Sovereign governments are responsible for health services to meet the needs of their citizens, and humanitarian organisations may only operate with the legal permission of the state. This places a tension between medical services operating under state authorities and those that operate under humanitarian principles. This paper postulates that health services that operate under the authority of a state cannot be neutral nor operationally independent. However, they must comply with the principles of humanity and impartiality of the Geneva Conventions and wider International Humanitarian Law. It considers the rights of protection for both military and civilian health services during conflict when the definition of combatant is blurred. It closes by discussing the implications for health services when at least one party deliberately targets non-combatants and other elements of critical civilian infrastructure as a means of waging war. Thus the phrase 'medical humanity' rather than 'medical neutrality' might be better to emphasise the right to protection for health services as non-combatant or civil defence organisations.

1 Introduction

'Medical neutrality' is a phrase used to emphasise a separation between healthcare and politics, particularly in the context of armed conflict and other emergencies(Orr and Fleming 2023). Conflict is inherently catastrophic to humanity both at an individual and societal level. Ever since societies organised to conduct war, efforts have been made to limit the consequences of war through laws, conventions, codes of

---

M. C. M. Bricknell (✉)
Centre for Military Ethics, King's College London, London, UK
e-mail: martin.bricknell@kcl.ac.uk

D. Messelken, A. E. Barbar (eds.), *Challenging Medical Neutrality*, Military and Humanitarian Health Ethics, https://doi.org/10.1007/978-3-031-69398-4_7

practice, and dialogue between combatants. The means and methods employed in the conduct of war are an indication of the values of the society represented by a party to war. The extension of wars from armies to societies and the increase in the destructive power of weapons during the nineteenth century led to the Lieber code for the Union Army during the American Civil War and the emergence of the Geneva and Hague conventions in Europe; both were an attempt to constrain the impact of war on humanity and to make provisions for the victims of war. The experiences of global conflict during the First and Second World War provided further impetus and consensus to develop agreements between States which would limit the destructive impact of war and protect those not involved in combat. There is now a substantial body of international law including the Geneva Conventions, collectively known as the Laws of Armed Conflict (LOAC), which proscribe the conduct of war by States and non-state armed groups (NSAGs). LOAC protects healthcare and medical services from attack as non-combatants.

In parallel, the international community has developed mechanisms for collaboration and coordination of humanitarian assistance to relieve the suffering of populations affected by conflict. Humanitarian relief during the late eighteenth century to World War 2 can be considered as an imperial act of social mobilisation (Barnett 2011). The international system of international relief originated from the United Nations Relief and Rehabilitation Administration that was set up during the Second World War to mitigate the impact of the destruction of Europe on civilian populations (Salvatici 2017). This led to an international consensus that enabled the United Nations (UN) to become a coordinator of international relief efforts during subsequent wars and other emergencies (O'Sullivan et al. 2016). The experiences and challenges associated with providing humanitarian relief during the Cold War resulted in the development of the Humanitarian Principles of humanity, neutrality, impartiality, and operational independence. A series of UN resolutions have empowered the UN to coordinate international humanitarian assistance between sovereign states and to organise provision of relief in emergencies where the authority of a state is non-existent. The humanitarian movement has sought to protect its activities and resources through a clear separation between their role and the violence committed by warring parties. This is justified both by the Humanitarian Principles and also through the application of many of the same laws covered by LOAC but using a different collective noun, International Humanitarian Law (IHL). The phrase *'medical neutrality'* invokes this lineage by implying healthcare is apolitical and humanity-orientated, solely focussed on the mitigation of the human consequences of war and protected under these laws and treaties.

Unfortunately there is evidence that the conduct of war by some States and armed groups is contrary to both LOAC and IHL. This is not new. During World War 2, the Geneva Conventions were, in general, followed at the tactical level in battle between the Allies and Germany, though they were completely disregarded by the Japanese in the Far East and by the Germans and Russians on the Eastern Front (MacNalty and Mellor 1968, Part 4.: The Geneva Conventions). More recently there is proof of attacks on healthcare facilities and health personnel in many areas of the conflict. Not only does this cause direct damage to medical facilities and personnel but it also undermines the willingness of health professionals to support populations exposed to violence because of the risk to themselves (Abu Sa'Da et al. 2013). The Healthcare in Danger initiative by the International Committee of the Red Cross (ICRC),[1] the 'Attacks on Medical Care' campaign by Medicins sans Frontiers,[2] and the development of several 'trackers' to record attacks on healthcare[3] have emerged over the course of the second decade of the twenty-first century to highlight this issue. The frequency and intensity of these attacks in some conflicts seem to suggest that these may be deliberate with an implication that there is culpable intent and purpose behind such transgressions of IHL (Haar et al. 2021). This must mean that in some wars, the attacker is deliberately targeting health facilities, and they believe that such attacks confer military advantage (Omar 2020). This chapter will examine the limitations of the concept of *'medical neutrality'* in the context of health services and whether the humanitarian principles apply to health services provided by a state rather than a humanitarian body. It will argue that the concept of *'medical neutrality'* in a theatre of war is wrong in the context of the Laws of Armed Conflict, and that *'medical humanity'* might be better to emphasise the rights of protection for all healthcare organisations based on their roles and identity as non-combatants or civilian objects whether they are military, state-provided, or humanitarian.

2 Definitions

The phrase 'medical neutrality' has emerged as a shorthand to cover the principles of freedom of access to healthcare for populations affected by conflict, and non-interference by armed actors with the provision of health services (Carruth 2020). It builds upon ideas enshrined in IHL that the clinical space in which healthcare is delivered should be protected from violence so that timely health services can be provided solely on the basis of clinical need (Benton and Atshan 2016). However, even access to healthcare can be a political, i.e. non-neutral, act if political actors seek to discriminate access to healthcare on non-clinical grounds (Orr and Fleming 2023). Thus the logic of phrase needs to be tested both through the lens of IHL to determine if medical care fits within the Humanitarian Principles and the lens of

[1] Marking Ten Years of Health Care in Danger. (2022) International Committee of the Red Cross, Geneva. Available at: https://www.icrc.org/sites/default/files/wysiwyg/Activities/Health/HCiD/4673_002_ten-years-of-hcid_web-1.pdf

[2] When medical care comes under attack. Medicins sans Frontiers. Available at: https://www.msf.org/attacks-medical-care-depth accessed 10 Jul 2023.

[3] Surveillance System for Attacks on Healthcare. World Health Organisation. Available at: https://extranet.who.int/ssa/Index.aspx; Attacked and Threatened: Health Care at Risk. Insecurity Insight. Available at: https://map.insecurityinsight.org/health; Illegal Attacks on Health Care in Syria. Physicians for Human Rights. Available at: https://syriamap.phr.org/#/en,

LoAC to determine if armed actors are obliged to recognise the 'neutrality of medical care'.

The Laws of Armed Conflict (LoAC) tend to be separated between 'jus ad bellum', the law under which States may resort to war to resolve international relations, and 'jus in bello', the law that governs the use of force during conflict. The debate covered by this chapter relates to 'jus in bello' and the contrasting perspectives and definitions used by armed forces compared to international humanitarian organisations. It will start by considering the provision of healthcare during conflict as an act of *'humanity'* under IHL based on definitions from a humanitarian perspective. It will then discuss how healthcare should be protected by armed actors using definitions from a LoAC perspective.

The UN General Assembly Resolution 46/182 of 1991 is considered to have created the international humanitarian system. It established the 3 humanitarian principles of *humanity*, *neutrality*, and *impartiality*.[4] It also emphasised the primary responsibility of States in meeting the needs of victims of emergencies and the duty of the UN humanitarian response to respect the sovereign authority of States. Subsequently, the concept of the 'complex emergency' was developed to reflect the challenge of delivering humanitarian assistance in conditions where civilian authority had broken down and there were high security risks for humanitarian relief workers.[5] UN General Assembly Resolution 58/114 of 2004[6] reaffirmed these principles and added the concept of *operational independence* to protect the autonomy of humanitarian objectives from the political, economic, military, or other objectives that any actor may hold in areas where humanitarian relief is being implemented. UN Security Council Resolution 1894 of 2009 further reaffirmed these 4 principles and the obligation of parties to armed conflict to comply with international humanitarian, human rights, and refugee law for the protection of civilians.

The UN Security Council passed a specific resolution in 2016 that condemned attacks on healthcare and called upon parties to armed conflicts to comply with IHL, especially the Geneva Conventions, due to the mounting evidence of such practice.[7] Table 7.1 shows the UN Office for the Coordination of Humanitarian Assistance (UNOCHA) definitions of these principles. Whilst these documents define the

[4] UN General Assembly Resolution 46/182. (1991) Strengthening of the coordination of humanitarian emergency assistance of the United Nations. United Nations, New York. Available at: https://undocs.org/Home/Mobile?FinalSymbol=A%2FRES%2F46%2F182&Language=E&DeviceType=Desktop&LangRequested=False

[5] Inter-Agency Standing Committee Working Group XVIth Meeting. 30 November 1994. UN OCHA, New York. Available at: https://interagencystandingcommittee.org/system/files/legacy_files/WG16_4.pdf

[6] UN General Assembly Resolution 58/114. (2004). Strengthening of the coordination of humanitarian emergency assistance of the United Nations. United Nations, New York. Available at: https://undocs.org/Home/Mobile?FinalSymbol=A%2FRES%2F58%2F114&Language=E&DeviceType=Desktop&LangRequested=False Accessed 10 July 2023.

[7] UN Security Council Resolution 2886/16. (2016). United Nations, New York. Available at: https://digitallibrary.un.org/record/827916?ln=en Accessed 10 July 2023.

Table 7.1 Definitions of humanitarian principles

Humanity—Human suffering must be addressed wherever it is found. The purpose of humanitarian action is to protect life and health and ensure respect for human beings
Neutrality—Humanitarian actors must not take sides in hostilities or engage in controversies of a political, racial, religious or ideological nature
Impartiality—Humanitarian action must be carried out on the basis of need alone, giving priority to the most urgent cases of distress and making no distinctions on the basis of nationality, race, gender, religious belief, class or political opinions
Independence—Humanitarian action must be autonomous from the political, economic, military or other objectives that any actor may hold with regard to areas where humanitarian action is being implemented

Source: OCHA on message—Humanitarian Principles. (July 2022). Office for the Coordination of Humanitarian Affairs. Available at: https://www.icvanetwork.org/uploads/2023/06/OOM_Humanitarian-Principles_Eng.pdf

international humanitarian system, none of them use the unique phrase 'medical neutrality'.

These principles have been used by UN agencies and international humanitarian organisations to emphasise their separation from political organisations and armed actors. They are embedded in the Humanitarian Charter which is a statement of shared beliefs for humanitarian organisations that may deploy in times of armed conflict or disaster to provide assistance to affected populations.[8] This Charter emphasises the non-partisan and independence of humanitarian organisations while recognising the primary responsibility of states to meet the needs of their citizens. Thus there is a tension between the presumed independence of humanitarian organisations and the government, as the executive authority of the state, in the provision of essential services to its population.

Moreover, the definition of the Humanitarian Principles blurs their application between *humanity* as a noun that can be applied to the moral behaviour of an individual of an organisation, *humanitarian* as an adjective to describe an activity that brings assistance or relief to a population, and *humanitarian* as an adjective which describes the principles under which an organisation might operate. The term *humanity* is also a tenet of LOAC based on the adjective *humane* which aims to constrain the application of violence by armed actors in order to limit the effects of war on both combatants and non-combatants in contrast with the adjective *inhumane*, meaning cruel or causing unnecessary suffering.

Whilst being violent and destructive, war differs from crime or violent acts by humans in that it is a political act under conscious control of an organised society. Therefore the conduct of war reflects the values and moral codes of the participants. The evolution of the character of war over the 20th and 21st centuries reflects the increasing span of domains or environments in which war might be fought (land, air, sea, space, and electromagnetic) and the increasingly destructive power of

[8] Humanitarian Charter. Sphere Movement. Available at: https://www.spherestandards.org/humanitarian-standards/humanitarian-charter/ accessed 10 July 2023.

military technologies to harm humans and their environment. These changes in the character of war have been reflected by attempts, especially after a conflict, to limit the impact of war through treaties covering the means of warfare such as chemical warfare, nuclear warfare, biological warfare, and landmines.

International humanitarian organisations argue that the collective phrase 'International Humanitarian Law' (IHL) serves to constrain the use of force during war in order to protect civilians and non-combatants. This contrasts with the collective phrase, the 'Laws of Armed Conflict' (LOAC), that tends to be used by State armed forces to authorise the use of force against their enemies. Both concepts reference the same body of treaties and customary law, particularly the Geneva Conventions and Associated Protocols (GCs), using the same terms but the interpretation of meaning have some significant differences. The ICRC Dictionary of the International Law of Armed Conflict defines *humanity* from an ICRC perspective as 'being sensible of and sharing the suffering of others, and preventing and alleviating it'. This and *impartiality*, as the 'positive principle of relief without discrimination', are similar to the humanitarian definitions (Verri et al. 1992). However in LOAC, *neutrality* primarily refers to the status of States that are not a party to armed conflict rather than organisations involved in humanitarian relief.

There is not a definition of *humanity* within LOAC, though the UK states that humanity 'forbids the infliction of suffering, injury, or destruction not actually necessary for the accomplishment of legitimate military purposes'.[9] It is strongly linked to concept of crimes against humanity within the definitions of war crimes contained within the Rome Statute of the International Criminal Court.[10] Thus, within the LOAC perspective, the principle of *humanity* applies to the nature of the use of force, alongside the principle of *military necessity*, the justification of the use of force solely to achieve war aims. This is balanced by *proportionality*, which requires that the means and methods of warfare should not be disproportionate to the military advantage sought. Finally, armed forces must apply *distinction* in their use of force between the armed forces and civilians, or between combatants and non-combatants, and between objects that might legitimately be attacked and those that are protected from attack. The concept of *distinction* includes the definition of non-combatant as a person who does not take part in hostilities and thus is protected from attack such as medical personnel, personnel assigned to medical duties, chaplains, and the wounded and sick (hors de combat). In war between States, the Geneva Conventions also place clear obligations upon States to ensure access to healthcare for the sick and injured without discrimination. This is different from the 'complex emergency' in which there is insufficient authority of States to meet these obligations. Both the traditions of IHL and LOAC assert the legal and ethical right of protection for non-combatants, civilians, and organisations providing relief to

[9] JSP 383 Manual of the Law of Armed Conflict. (2014) Ministry of Defence, London. Available at: https://www.gov.uk/government/collections/jsp-383

[10] Rome Statute of the International Criminal Court Article 8. (2011) ICC. The Hague. Available at: https://www.icc-cpi.int/sites/default/files/RS-Eng.pdf

populations affected by armed conflict or disaster. Neither use the phrase *'medical neutrality'*, however medical services meet all interpretations of the word *humanity*.

In spite of IHL/LOAC, states and non-state armed groups (NSAGs) may not be signatories to the primary law or conventions, or may choose not to apply them in their conduct of war. Individual fighters or tactical leaders may interpret 'military necessity' as an authorisation for the unrestrained use of force. Even if there is prima facie evidence of a war crime, there are limits to the application of consequences of contraventions of such law if the state is a party to the violence or the state does not seek to investigate or punish perpetrators. Hugo Slim, in his book Solferino 21, challenges the 'western, international system' of humanitarian aid in light of the emergence of Chinese, Russian, Indian and Middle Eastern state and non-state actors in the provision of humanitarian relief (Slim 2022). He argues for more localisation and more commitment to delivery through state organisations. He also suggests that political neutrality is not a sine qua none for humanitarianism, especially in situations where state authorities are explicitly discriminating against defined populations. Rubenstein, reviewing the struggle to protect healthcare from the violence of war, argues that healthcare has always been vulnerable to attack during armed conflict. He suggests that reducing the impunity from punishment for perpetrators of war crimes through gathering evidence, lobbying for respect of IHL/LOAC, and demonstrating the value of impartial provision of healthcare are the methods to reduce attacks on healthcare (Rubenstein 2021). Both argue that humanitarian assistance might have a political dimension and that health workers are not obliged to be neutral though they must retain their ethical obligations to be humane and impartial.

This section has argued that the phrase *'medical neutrality'* does not have provenance within IHL or LOAC. The sick and injured are entitled to access to healthcare, health services are to be protected, and IHL and LOAC places duties on parties to conflict to comply with these rights. The Humanitarian Principles and the Humanitarian Charter provide guidance for the behaviour of international humanitarian organisations but compete with the sovereignty and duties of the state to meet the needs of its own citizens. The next section considers whether the civilian services (including health) of a state can be neutral if it is a party to armed conflict.

3 Can the Services of the State Be Neutral?

Both IHL and LOAC establish a clear distinction between the State as a warring party and the freedom from attack to be afforded to States, or other international parties, that declare themselves to be neutral. Thus a State 'at war' cannot be neutral and the services of a State cannot be neutral as they are under the direction of a warring party. A State's 'combatants' are legitimate targets. In addition to directing the employment of combatant forces, the State directs the other entities of the state that support war, most particularly industrial production and logistics. This was the justification for attacks on these capabilities by the Allies in Germany and Japan during the Second World War and by the USA in North Vietnam, Cambodia, and Laos during the Vietnam War. The application of *distinction* between a combatant and the corollary, non-combatant (including civilians), has become even more difficult in the context of non-kinetic means to achieve military advantage by 'combat' in the space and cyber domains, and the use of weapons which may be controlled remotely such as drones. Thus the definition of a legitimate target (military objective) might be determined by the function of a facility rather than the uniforms or other distinguishing features of the individuals employed within it; in the same way that militias or resistance movements can be defined as combatants (and have rights and duties under LOAC/IHL). Thus the application of '*military necessity*' by an attacker can be more problematic once such attacks are directed away from identifiable combat forces, and ultimately are justified in the perceptions of the victor.

LOAC defines various categories of a State's assets, people and infrastructure, as protected and should not be the subject of attack. The definition of *non-combatants* contrasts with the definition of *combatant* members of the armed forces and covers medical personnel, personnel assigned to medical duties, and chaplains. *Non-combatants* have duties to the provide medical and spiritual care to victims of conflict, and rights of protection using specific identity documents and the Geneva emblems. A civilian is defined as a person who is not a combatant, namely not a member of the armed forces, militia or volunteer corps. Other groups specifically defined as *protected* from attack include: wounded, sick, shipwrecked, prisoners of war, parlementaires (truce negotiators), civil defence personnel, and the previous groups of civilians, medical personnel and religious personnel of the armed forces. Civilian objects, which should not be attacked, include: buildings for religion, education, art, science or charitable purposes, historic monuments, hospitals and places where the sick and wounded are collected, and objects indispensable to the survival of civilians such as foodstuffs, livestock, power and water.

Based on the previous arguments, a State cannot be neutral, personnel not wearing military uniform can be combatants if working in support of the military functions of the State, and yet the State has primary responsibility to meet the needs of victims of humanitarian emergencies within its own borders. Therefore the humanitarian principle of *neutrality* cannot apply to the functions of the State even if the purpose of instruments of the State is the provision of humanitarian assistance to national populations in distress. This clearly applies to health services and so state-provided health services cannot claim *'medical neutrality'* as they are provided by the state and subject to national law. Under both LOAC and IHL, humanitarian relief provided by the State derives protection from attack on the basis that they are not conferring military advantage to the State, are undertaking the function of *humanity*, and so are *non-combatants* or are *civil defence personnel*. This also applies to non-state/non-government organisations that are supporting the provision of government services such as healthcare. LOAC and IHL also mandate governments to facilitate access to health services purely on the basis of clinical need and without discrimination, which lies within the phrase *'medical humanity'*. Thus the phrase *'medical neutrality'* does not confer any additional rights or protection for health services beyond those already covered by both LOAC and IHL.

4 Protection of Health Services in War

However, if one considers that the purpose of war is to achieve political outcomes after the exhaustion of other means, and the authority of the government of a State is based upon a mandate from the people of the state, it is inevitable that warring parties will consider 'the will of the people' as the military centre of gravity for both themselves and their adversaries. Therefore it cannot be assumed that customary international law or treaties will necessarily translate into protection from attack. Thus the State has to consider how to balance legal protection with other means to protect their population and its assets.

Identification of non-military objects would seem to be the primary means of protection as this is a pre-requisite for the employment of *distinction* by combatants. Once a non-military object has been identified, the combatant may not attack it under 'jus in bello'. The legal basis for identification for medical personnel, facilities or transports is founded on the use of the Geneva Emblems, the Red Cross, Red Crescent, Red Lion or Red Crystal. These can be worn as armbands or tabards by individuals and applied to medical transports or facilities. However, they need to be visible to the enemy to achieve their purpose. This might be difficult at night, at a long distance, in bad weather, or in the presence of battlefield smoke. This challenge of identification was already recognised prior to the First World War (Donegan 1904). In the inter-war years there was discussion about the practical expectations of the Red Cross for protection against aerial bombardment and war gases (Cowell 1939). Military objects should be sited away from protected objects and vice-versa. It has also been suggested that transponders or other methods of identification in the electromagnetic spectrum may help to identify protected entities. Exchanges of location data for hospitals and other protected sites through neutral third parties have also been used to aid identification; and misused to enable deliberate attacks on health facilities.[11]

If the adversary does not seem to comply with the LOAC then all such measures will fail, and the best chances of protection are from disguise and physical measures. Whilst there may be a separation between military and civilian health systems in peace, in war or other health emergencies the entire national health economy will be mobilised to meet the health needs of the population, including the armed forces. The military medical services will focus on field medical care for the armed forces but are also required to provide immediate treatment for all casualties that they receive without discrimination, except on medical grounds. It is likely that the active-duty component of the military health system will be augmented by Reserves and volunteers from the civilian health system. The civilian health system will also

[11] Letter dated 6 April 2020 from the Secretary-General addressed to the President of the Security Council. United Nations Headquarters Board of Inquiry into certain incidents in the north-west of the Syrian Arab Republic since 17 September 2018 involving facilities on the United Nations deconfliction list and United Nations-supported facilities. Available at: https://undocs.org/S/2020/278 Accessed 14 Jul 2023.

care for all patients that present to their facilities, including military personnel. The national health system may be augmented by charities, the commercial sector, or international humanitarian organisations. Civilians may be present in an area of active operations, especially in urban areas, and so military and civilian health services may work together in close coordination. As seen during the COVID-19 crisis, this co-ordination is likely to be replicated through the entire hierarchy of the national health system. The care pathway will include the rehabilitation and recovery of members of the armed forces to enable many to return to duty.

Therefore the health system within a State at war cannot be neutral nor operationally independent of the State, and thus cannot meet the principles of a humanitarian organisation. However it serves a humanitarian function and is legally and ethically obliged to ensure equitable access and treatment to all patients. Whilst protected under LOAC and IHL, the health system is at risk of attack either as a deliberate act or as inadvertent collateral damage from an attack on a military object. Therefore it is necessary to take appropriate and proportional measures to protect and mitigate against the consequences of an attack. This might include physical protective measures such as moving into basements and underground bunkers, sandbagging, dispersal of services, cyber-protection, and active collective defence such as air defence zones. It will also be necessary to mitigate against the effects of attack through backup services for water, power, oxygen, communications, and medical records. The State should also consider tracking any attacks on healthcare and collecting contemporary evidence of attacks for any subsequent investigation or criminal inquiry.

5 Conclusion

In conclusion, access to and provision of healthcare for victims of conflict is a humane act enshrined in the sovereign duty of States within the Laws of Armed Conflict and International Humanitarian Law. These rights and duties might be summarised by the phrase *'medical humanity'*. The provision of health services by a State as war cannot be a neutral act nor can health services be operationally independent of the State. The State is likely to co-ordinate the actions of all providers of health services including military, civilian and humanitarian to meet the needs of its population. Thus the phrase *'medical neutrality'* does not encompass the activities of all health providers and can exclude the provision of medical care by military or politically active organisations. Whilst protected under LOAC and IHL, health services may be deliberately or inadvertently attacked during war. Therefore health services should take active and passive measures to mitigate the threat and consequences of such an attack. It is a fallacy to use the phrase *'medical neutrality'* in a country at war and to expect State health services to comply with all of the humanitarian principles. However the value of IHL/LOAC remains undiminished as an expression of humanity during war as the most inhumane of human endeavours. This is best summarised by the final paragraph of the final chapter on the Geneva

Conventions from the Principal Medical Lessons of Second World War (MacNalty and Mellor 1968, Part VI. The Geneva Conventions).

The Red Cross is a symbol of man's compassion and those whose task it is to work under its protection take on a truly positive and beneficent role in warfare. The cynic might suggest that the increasing power and speed of engines of war will nullify all their endeavours. That is not so. In war there has grown up a strong adherence of participants to a pattern of conduct. International Law, even if not always codified, has great force and weight and goes much of the way in setting moral as well as legal standards. Despite time and change the Geneva Cross and all it stands for will always be a powerful help to man in time of war.

References

Abu Sa'Da, Caroline, Françoise Duroch, and Bertrand Taithe. 2013. Attacks on medical missions: Overview of a polymorphous reality: The case of Médecins Sans Frontières. *International Review of the Red Cross* 95: 309–330. https://doi.org/10.1017/S1816383114000186.

Barnett, Michael. 2011. *Empire of humanity*. Cornell University Press.

Benton, Adia, and Sa'ed Atshan. 2016. "Even war has rules": On medical neutrality and legitimate non-violence. *Culture, Medicine, and Psychiatry* 40: 151–158. https://doi.org/10.1007/s11013-016-9491-x.

Carruth, Lauren. 2020. Medical neutrality. In *Humanitarianism: Keywords*, ed. Antonio De Lauri, 128–130. Leiden Boston: Brill.

Cowell, E.M. 1939. The protection of medical establishments by signs. *Journal of the Royal Army Medical Corps*: 289–293.

Donegan, J.D.F. 1904. The Geneva convention in modern warfare. *Journal of the Royal Army Medical Corps* 2: 12–19.

Haar, Rohini J., Róisín Read, Larissa Fast, Karl Blanchet, Stephanie Rinaldi, Bertrand Taithe, Christina Wille, and Leonard S. Rubenstein. 2021. Violence against healthcare in conflict: A systematic review of the literature and agenda for future research. *Conflict and Health* 15: 37. https://doi.org/10.1186/s13031-021-00372-7.

MacNalty, Arthur Salusbury, and W. Franklin Mellor. 1968. *Medical services in war: The principal medical lessons of the second world war*. Vol. 1. London: HM Stationery Office.

Omar, Abdulaziz. 2020. Understanding and preventing attacks on health facilities during armed conflict in Syria. *Risk Management and Healthcare Policy* 13: 191–203. https://doi.org/10.2147/RMHP.S237256.

Orr, Zvika, and Mark D. Fleming. 2023. Medical neutrality and structural competency in conflict zones: Israeli healthcare professionals' reaction to political violence. *Global Public Health* 18: 2171087. https://doi.org/10.1080/17441692.2023.2171087.

O'Sullivan, Kevin, Matthew Hilton, and Juliano Fiori. 2016. Humanitarianisms in context. *European Review of History: Revue européenne d'histoire* 23: 1–15. https://doi.org/10.1080/13507486.2015.1117422.

Rubenstein, Leonard S. 2021. *Perilous medicine: The struggle to protect health care from the violence of war*. New York: Columbia University Press.

Salvatici, Silvia. 2017. 'Fighters without guns': Humanitarianism and military action in the aftermath of the second world war. European Review of History: Revue européenne d'histoire: 1–20. https://doi.org/10.1080/13507486.2017.1374354, 1, 20.

Slim, Hugo. 2022. *Solferino 21: Warfare, civilians and humanitarians in the twenty-first century*. London: Hurst & Company.

Verri, Pietro, Edward Markee, and Pietro Verri. 1992. *Dictionary of the international law of armed conflict*. Geneva: International Committee of the Red Cross.

Chapter 8
Medical Neutrality and Impartiality in UN Peace Keeping Operations

Min Yu and Yuxuan Yang

Abstract Medical professionals have been playing an important and unique role in United Nations' (UN) Peacekeeping Operations. They come from different countries around the world, serving people with diverse religious, cultural, and educational backgrounds. To guide their activities, the UN formulated and distributed the Medical Support Manual, including the ethical code for medical practitioners. Furthermore, in addition to their national codes of ethics, during the services in the United Nations, all physicians and United Nations staff or volunteer physicians, shall be guided by and consistent with the International Code of Ethics for Occupational Health Professionals and the International Code of Medical Ethics. As peacekeepers, medical professionals follow three basic principles: consent of the parties, impartiality, and non-use of force, which continue to set UN peacekeeping operations apart as a tool for maintaining international peace and security for decades. Impartiality had been a key code for UN medical professionals' practices but was misunderstood for over four decades until 2000. Years ago, the UN made a clear definition of "impartiality," which was considered crucial to maintaining the consent and cooperation of the main parties. With this change, UN peacekeepers should be impartial in their dealings with the parties to the conflict but not neutral in the execution of their mandate. Affected by this change, all the peacekeepers have been facing new challenges. Some cases are discussed to analyze the dilemma of medical professionals for peacekeeping missions in this chapter.

1 Introduction

United Nations' (UN) peacekeeping operations (PKO) have over 80 years of history and gained a reputation for maintaining peace and protecting civilians suffering from conflicts. Military personnel have always been the backbone of peacekeeping forces, however, due to the differences between peacekeeping missions and

M. Yu (✉) · Y. Yang
Academy of Military Medical Sciences, Beijing, China

D. Messelken, A. E. Barbar (eds.), *Challenging Medical Neutrality*, Military and Humanitarian Health Ethics, https://doi.org/10.1007/978-3-031-69398-4_8

traditional military tasks, they face new ethical challenges. To help these uniformed peacekeepers overcome challenges, the UN developed three basic principles for PKO: consent of the parties, impartiality, and non-use of force.

2 Challenges Faced by United Nations' Peacekeeping Operations Medical Services

2.1 History of UN PKO

The idea of UN Peacekeeping Operation first appeared during Cold War, and the first UN Peacekeeping Operation UNTSO (United Nations Truce Supervision Organization) happened in 1948 when the UN deployed military observers to the Middle East to monitor the Armistice Agreement between Israel and its Arab neighbors ("Our History UN Peacekeeping," n.d.). At that time, UN PKO's mission was limited to maintaining ceasefire, stabilizing the local situation, and providing crucial support for political efforts to resolve conflict. Only unarmed military observers and lightly armed troops were involved in PKO with the duty of monitoring, reporting, and confidence-building. Between 1948 and 1988, a total of 15 UN PKOs was authorized by the Security Council ("List of Peacekeeping Operations," n.d.). Over 20,000 peacekeepers were deployed to mission areas, and 794 sacrificed their lives during missions ("Fatalities by Year," n.d.).

After the Cold War, the missions of UN PKO expanded from traditional observational tasks performed by military personnel to multidimensional missions involving ensuring the implementation of comprehensive peace agreements and assisting in laying the foundations for sustainable peace. Over time, the nature of conflicts shifted from interstate to intrastate conflicts, and civil wars and tasks required undertaken by the peacekeepers became more and more complex. Therefore, as the military remained the major force of PKOs, more and more legal experts, human rights monitors, and humanitarian workers became peacekeepers. There was a rapid increase in the number of PKOs. However, starting in the mid-1990s, the UN PKO faced challenges of establishing missions that were hard to accomplish in areas like Somalia and former Yugoslavia, where no peace was there to keep. Peacekeepers serving these missions were also performing "peacebuilding and peacemaking tasks" of which conflicts were more intense and peacekeepers' safety were at high risk. During the 1990s, a total of 38 PKOs were authorized ("List of Peacekeeping Operations," n.d.), with more than 75,000 peacekeepers deployed to mission areas, and 824 of them died while serving missions ("Fatalities by Year," n.d.). Among these 824 fatalities, 30.58% occurred during the UN Operation in Somalia II (UNOSOM II), and 20.27% occurred during UN Protection Force (UNPROFOR) in former Yugoslavia ("Fatalities by Mission And appointment Type," n.d.). These numbers indicated that UN PKO at that time did not have enough capacity and financial support to deal with situations where guns had not yet fallen silent. Therefore, reflection and improvement were needed for the UN to keep up with the increasingly difficult and complex PKO demands worldwide.

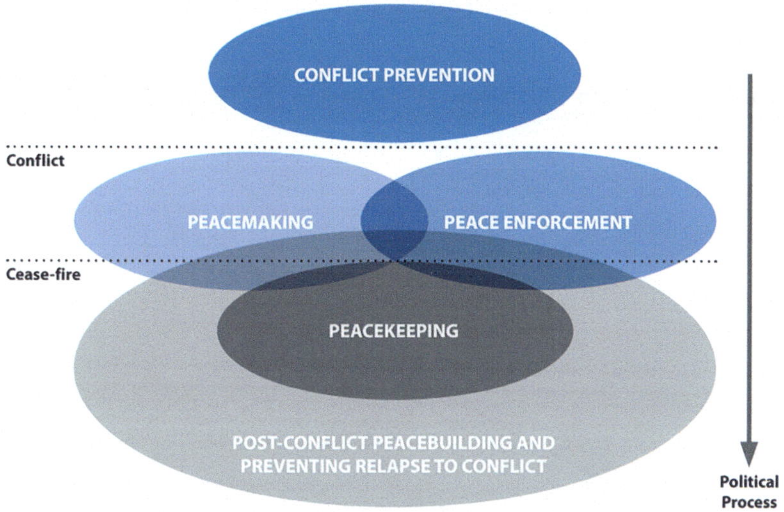

Fig. 8.1 Relationship between conflict prevention, peacemaking, peace enforcement and peacekeeping

Entering the twenty-first century, the UN first reexamined the existing PKO system, found shortcomings, and provided actionable recommendations; the results were written in the Brahimi Report underscoring the importance of securing adequate resources and operating under a clear and achievable mandate. Following the Brahimi report, the UN published Capstone Doctrine in 2008 to guide all peacekeepers regarding the most important principles, among which clarifying the definition of peacekeeping and its relation to conflict prevention, peacemaking, peace enforcement, and peacebuilding ("United Nations Peacekeeping Operations Principles and Guidelines," 2008). It underscored those peacekeeping missions should happen under ceasefire situations, but that may not be the case in reality and there are linkages and grey areas among these peace related activities (Fig. 8.1).

As of May 2023, there have been 71 PKO missions deployed by the UN, peacekeepers from over 120 countries have participated in those operations, and 4314 peacekeepers died during peacekeeping operations ("Our History UN Peacekeeping," n.d.).

2.2 *Current UN PKO*

UN PKO is now entering a consolidation stage as the number of new operations slightly declined after substantial growth and a surge in new operations. Despite the fact that the complete number of missions diminishes, the demand for PKO field missions will stay high because of the development of new conflicts spreading beyond local and regional boundaries. Currently, there are 12 active peacekeeping

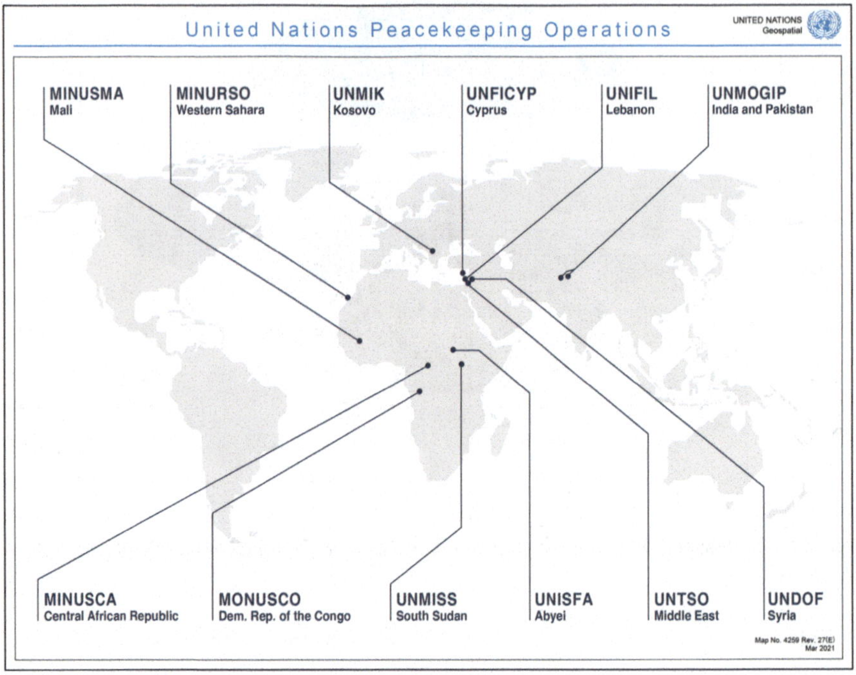

Fig. 8.2 Active United Nations peacekeeping operations (May 2023)

operations (Fig. 8.2) worldwide, with 88,681 personnel from 121 countries participating in those missions. Among those personnel, 85.01% (75,384) are military personnel, indicating that the military remains the backbone of UN PKOs ("Peacekeeping Operations Fact Sheet," 2023).

Nowadays, multidimensional peacekeeping operations focus on the following missions:

1. Prevent conflicts.
2. Protect civilians.
3. Build rule of law and security institutions.
4. Advance political solutions to conflict.
5. Protect and promote human rights.
6. Promote women, peace and security.
7. Promote youth, peace and security.
8. Deliver operational support.

According to the Action for Peacekeeping (A4P) initiative, several challenges UN PKO faces today include the increasing threat to UN peacekeepers' lives, which causes a rise in casualties and, in some situations, lacking medical capacities and professionals may directly cause peacekeepers' deaths ("Action for Peacekeeping," n.d.).

2.3 Challenges Faced by Medical Support

The most important role of UN PKO medical support is to help deployed personnel maintain well-being in terms of both physical and mental. Besides, some also provide humanitarian aid to local civilians. Quality health service is defined as "planning, coordination, execution, monitoring, and professional supervision impacts the health and well-being of mission members and ultimately the mission's successful execution (Fig. 8.3).

UN PKOs are increasingly involved in dangerous areas and local parties of conflicts may disregard peacekeepers' security and role. A statistically significant and positive relationship exists between a mission's scale, remoteness, complexity, and aggressiveness and a higher probability of wounds or death, especially as a result of hostile attacks or diseases. The UN peacekeeping operations are hoped

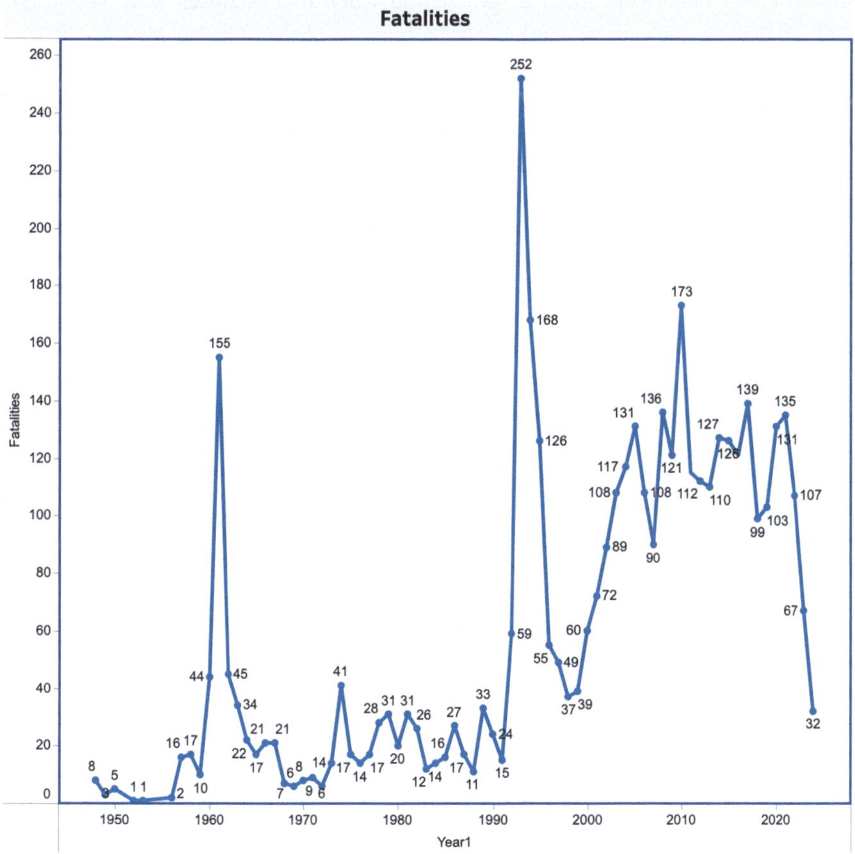

Fig. 8.3 Fatalities of UN peacekeeping operations since 1948

to become more streamlined, faster, more efficient, and more capable. There is no doubt that medical support will conform to the larger structure of UN peacekeeping operations.

From January 2018 to February 2019, there were 230 hostile acts targeting the UN peacekeeper, among which 67 of them resulting in casualties. There were 219 peacekeepers injured during those acts, and 40 of them died. UN peacekeepers serve at great personal risk to protect civilians, and medical services play a very important role in treating peacekeepers facing threats. Below are two real examples of UN peacekeepers being the target of hostile attacks.

On May 31, 2016, in Mali, a Chinese engineering camp was attacked by a suicide bomb. Sixteen casualties arrived at the Chinese Level II hospital. Two of them were seriously injured. One was diagnosed with an open fracture of the right lower leg and soft tissue defect of the left thigh, and the other was diagnosed with shock, facial soft tissue laceration, facial nerve injury, left ear injury, left shoulder bruise, and left ankle injury. Among these casualties, one received major emergency trauma surgery, two debridement suture, and ten debridement treatment. The two seriously injured cases underwent medical evacuation when their statuses were stable (Jia et al. 2018).

On July 10, 2016, as the crossfire continued in the urban area of Juba, the capital of South Sudan. A detachment of the Chinese peacekeeping infantry battalion was suddenly attacked during the execution of a United Nations peacekeeping mission resulting in the death of one person and the injury of 6 people, of which three were seriously injured and three were slightly injured ("Juba crisis–July 10," 2016).

Effective medical support is essential in enabling the peacekeepers deployed in complex peace missions to fulfill their mandates. It is also important for the health and security of deployed peacekeepers and for the operation's capacity to provide support to other players. Furthermore, the pattern in which medical support is delivered can affect the local authority. It can either support the peacekeeping operation's objectives or destroy them.

The new challenges of the complex UN peacekeeping operations are characterized by multidimensional mandates, multinational participation, rapid deployment, and volatile/hostile operational environments. To overcome these challenges, the corresponding countermeasures include: setting up an international capability development standard for medical contingents and creating a roster of medical experts potentially available to the mission start-up teams. Only once the possibility of a peacekeeping operation arises would they conduct the technical assessment and deploy for mission start-up to gather information about the host nation's available support. Training standby troops in the contributing countries and working with countries that maintain natural disaster relief capacities or other crisis means can help to improve mechanisms for states to make their resources available to the UN for mission start-up (Yu et al. 2018).

3 Ethical Codes for UN Medical Practitioners

3.1 Code of Conduct for All UN PKO Personnel

Since 1998, uniformed peacekeeping personnel have been provided with pocket cards of the *Ten Rules: Code of Personal Conduct for Blue Helmets* ("Ten Rules: Code of Personal Conduct for Blue Helmets," 1995, Fig. 8.4). In these ten rules, peacekeepers are asked to "display the highest integrity and impartiality" and "respect and regard human rights of all".

3.2 Ethics Code Specifically for UN PKO Medical Professionals

According to the first edition of the Medical Support Manual for United Nations Peacekeeping Operation (United Nations Department of Peacekeeping Operations 1995) published in 1995, the word "ethics" only appeared once in Chapter 5 as follows:

> The provision of medical care to the local population is a sensitive matter and must be weighed against humanitarian principles and the ethical code of medical practice.

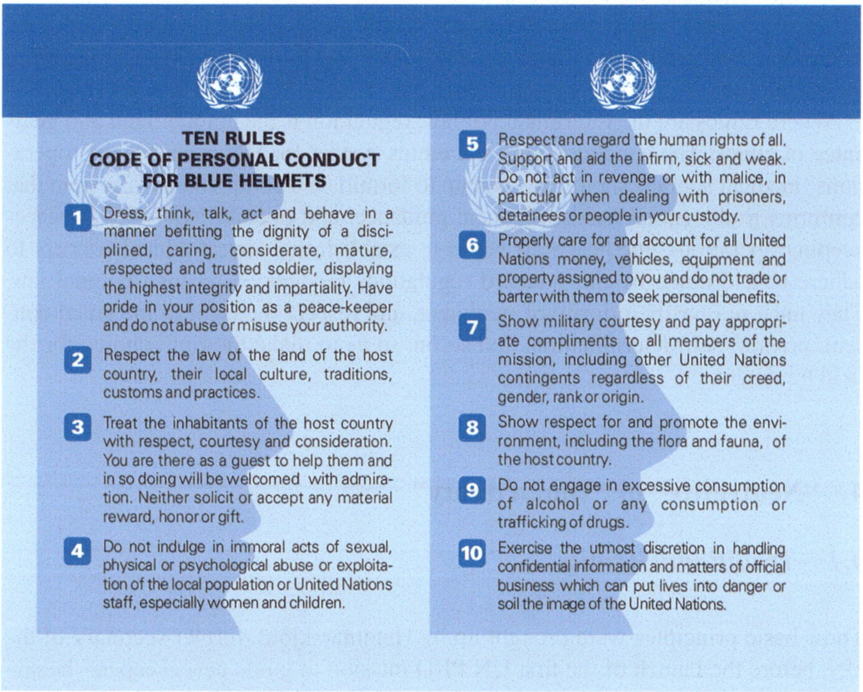

Fig. 8.4 Ten rules: code of personal conduct for blue helmets

No further explanation was provided to help guide medical professionals' behavior. It's clear that ethics was not being paid enough attention at that time. In 1999, the situation didn't get better as the second edition of the medical support manual got published, and content related to ethics remained the same sentence (United Nations Department of Peacekeeping Operations 1999). Military medical support experts, as well as humanitarian law experts, realized this problem and added one chapter to the updated third edition of the medical support manual (United Nations Department of Peacekeeping Operations and Department of Field Support 2015). Chapter 8 of the latest edition of the medical support manual covers ethical codes for medical practitioners and principles of medical ethics, which are listed as follows. The manual requires all physicians, including Troop/Police Contributing Country provided physicians and United Nations staff or volunteer physicians, shall be guided by and consistent with the International Code of Ethics for Occupational Health Professionals and the International Code of Medical Ethics, in addition to their country's code of ethics.

Some scholars argue that there are only a few pages about ethics, mainly coming from International War Law and International Humanitarian Law, such as the Hague Convention, the Geneva Conventions, and the Additional Protocol on the field of ambulance-related provisions, dedicated to the general level of medical ethics (Yu et al. 2018). The mission does not have more regulations particular to the unique circumstances and tasks. Due to the uniqueness of UN PKOs and their difference from traditional military tasks, medical personnel are more likely to face ethical challenges related to dual loyalties, and they may often encounter moral dilemmas of to what extent they should remain neutral during peacekeeping missions. Therefore, more ethical guidance should be provided to help them better perform their jobs.

Where Ethics are the forerunners of law, regulation is the embodiment and guarantee of ethics. The construction of the ethics system for UN peacekeeping operations' medical support should be sped up to formulate a comprehensive system that conforms to the actual needs, practical guidance, and operability. The UN peacekeeping operations' medical staff needs to establish the correct ethical concept to adhere to and abide by the standard regulations recognized by international law. They must master the priority of treatment, and respect the right to informed consent, privacy, self-determination, and so on, so as to make the right choice for the field treatment.

4 "Neutrality" or "Impartiality"

4.1 Traditional Trinity

Three basic principles were brought up by Hammarskjöld, former secretary of the UN, before the launch of the first UN PKO mission to guide peacekeepers' behavior: consent of the parties; impartiality; and non-use of force except in self-defense and defense of the mandate. These basic principles are also known as the "traditional trinity" and are still valid in today's peacekeeping operations.

First, UN PKOs are authorized after gaining consent from the main parties to the conflict. If parties in a conflict deny the intervention of peacekeeping operations, the UN will not be able to perform mandated tasks due to the high political and physical risks of becoming a part of the conflict instead of being impartial, which disregards the fundamental role of keeping the peace. What's more, the main parties don't represent all the parties in the conflict since the main parties may have weak control at the local level. Therefore, UN peacekeepers are still facing high personal risks. However, it is hard for the UN to get all parties consent in such volatile situations; therefore, getting only the main parties' consent is the bottom line before authorizing any missions or dispatching any peacekeepers.

Second, in order to gain consent from the main parties and collaborate to maintain peace, it is important for peacekeepers to remain impartial. According to the UN website, "peacekeepers should be impartial in their dealings with the parties to the conflict, but not neutral in the execution of their mandate" ("United Nations Peacekeeping Operations Principles and Guidelines," 2008). Peacekeeping operations should not take sides with the conflict parties, but it is acceptable to penalize parties that violate the international norms or principles a PKO upholds.

Third, even though UN peacekeeping operations are not enforcement tools, peacekeepers are allowed to use force in terms of self-defense and defense of the mandate, and it has to be approved by the Security Council. As mentioned before, the safety of peacekeepers is at high risk during some missions, and using force may help them survive hostile situations. Besides self-defense, UN peacekeeping operations should only use force as a measure of last resort in certain volatile situations that PKOs "robust" mandates authorizing them to "use all necessary means" to deter forceful attempts to disrupt the political process, protect civilians under imminent threat of physical attack, or assist the national authorities in maintaining law and order.

4.2 Neutrality or Impartiality

The word "impartiality" has been written in the holy trinity for over eight decades but was misunderstood and used during the first four decades. In 1956, the assumption was impartiality simply meant the absence of partiality, and military medical personnel considered "impartiality" a synonym of "neutrality," and these two words were often used together. However, that was not the case, and there was a clear difference between these two words. Even though both "neutrality" and "impartiality" suggest offering no assistance to any party in a conflict, a neutral party involves no judgment or decided views, whereas an impartial party does have "fair and just" standpoints. Impartiality means the absence of favoritism, prejudice, or any kind of discrimination, but it allows actors to take positions unattached to their parties (Donald 2003).

At the beginning of UN PKOs, the basic principle of "impartiality" was actually the meaning of "neutrality," and the UN required peacekeepers not involve in any conflicts with no standpoints. This situation changed in 2000 when the UN conducted a comprehensive reflection on its previous PKO missions after the cold war, and it clarified the definition of "impartiality" in the Brahimi Report of 2000. In this report, the UN states that "local parties consist not of moral equals but of obvious aggressors and victims, and peacekeepers may not only be operationally justified in using force but morally compelled to do so" ("Brahimi Report," 2000) For example, the genocide that happened in Rwanda should have been stopped by the international community; however, the misunderstanding of the principle of impartiality resulted in the omission of peacekeeping parties to oppose the evil action.

4.3 *Cases from Australian Peacekeeping Experience*

It would be better if we could use real-world examples to analyze the differences and pros and cons of impartiality and neutrality. The two cases shown below may help understand the differences between neutrality and impartiality and may lay out some ethical problems involved in peacekeeping operations, and it is obvious that different options may cause different problems. The first case represents the "impartial action" of peacekeeping, in which overwhelming forces were used to suppress violence. The second case represent the "neutral action" of peacekeeping, in which peacekeepers are only observers that provide humanitarian aid.

4.3.1 Case 1

The first case happened in Somalia when Australian peacekeepers took control of the city of Baidoa. During a year of time, they suppressed local militaries and gunmen and trained a local police force, and provided medical services to local civilians. It seemed like helping improve the condition of the city. However, after the peacekeepers left Baidoa, the situation returned back to the same, if not even worse. As soon as the peacekeeping troop was withdrawn, the violence got back to the city since the underlying social problems were not solved. It was clear that the peacekeepers fulfilled their responsibilities of maintaining peace and providing aid in a short period of time, but the goal of achieving permanent peace seems hard to be achieved (Joseph 1995).

4.3.2 Case 2

The second case happened in Rwanda in 1994. After the civil war in Rwanda, the new government decided to disband refugee camps, and a team of Australian medics were there to provide aid to civilians. The action of disbanding a camp happened several times before and seldom involved violence. However, it was not the case in this refugee camp, and a massacre happened within several hours resulting deaths of over 4000 refugees (Blocq, 2006). The peacekeepers stood by the sidelines, with no force enforcement, just occasionally venturing forth to treat the wounded. The troops were entitled to intervene in that situation, but the consequences might be they would be overwhelmed by the greater firepower of the new government troops.

One way to analyze different approaches is to take a look at the consequences of military force enforcement: will it bring more good than harm to the local situation and civilians? Unfortunately, it is hard to answer this question. On the one hand, it is wrong to watch someone suffer or die if you have the power to intervene. On the other hand, the long-term effect of interventions is not clear, and peace is not that easy to keep. As a UN peacekeeper, we would say it is not right to stand by and watch evil actions like massacres happen. Therefore, it would be better to be impartial than neutral while serving peacekeeping missions since peacekeepers need to have a standpoint of fairness and justice to help them better serve their job of maintaining peace.

5 Conclusion

As conflicts happen from time to time around the world, UN peacekeeping operations remain to be one of the most complex operations in terms of maintaining peace, protecting civilians, and contributing to peacebuilding. Unlike traditional military tasks, military personnel, including medics, will face totally different working environments and ethical challenges. After going through the latest version of the medical support manual for UN PKO, medics participating in PKOs are not given enough ethical guidelines given the fact that they will face different situations from their traditional work environment. Therefore, it is suggested that more detailed guidelines involving case analysis should be added to the next edition of the medical support manual. In addition, impartiality seems to be more suitable to UN PKOs compared to neutrality since staying impartial but not neutral during missions allows them to stand up against obvious wrong actions (usually violent actions) taken by parties of conflicts and use enforcement forces to defend the core value of peacekeeping.

References

Action for Peacekeeping. n.d. United Nations. https://peacekeeping.un.org/en/action-for-peacekeeping-a4p. Accessed 22 Jul 2023.

Brahimi Report. 2000. United Nations. https://peacekeeping.un.org/sites/default/files/a_55_305_e_brahimi_report.pdf. Accessed 22 Jul 2023.

Blocq. 2006. The fog of UN peacekeeping: Ethical issues regarding the use of force to protect civilians in UN operations. *Journal of Military Ethics* 5: 201–213.

Fatalities by Mission and appointment Type. n.d. United Nations. https://peacekeeping.un.org/sites/default/files/stats_by_mission_appointment_type_3_88_june_2023.pdf. Accessed 22 Jul 2023.

Donald. 2003. Neutral is not impartial: The confusing legacy of traditional peace operations thinking. *Armed Forces & Society* 29: 415–448.

Fatalities by Year. n.d. United Nations. https://peacekeeping.un.org/en/fatalities. Accessed 22 Aug 2024.

Jia et al. 2018. Experience on mass casualty procedure after terrorism attack in MINUSMA Chinese level II hospital, Aisa-Pacific Congress on Military Medicine.

Joseph. The Ethics of Peacekeeping. 1995. http://aapae.org.au/documents/article1995_joseph.pdf#:~:text=THE%20ETHICS%20OF%20PEACEKEEPING%20Keith%20Joseph%20Traditionally%20the,force%20as%20an%20integral%20part%20of%20peacekeeping%20missions. Accessed 22 Jul 2023.

Juba crisis—July 10. 2016. United Nations. https://unmiss.unmissions.org/sites/default/files/160710_juba_crisis_-_july_10.pdf. Accessed 22 Jul 2023.

List of Peacekeeping Operations. n.d. United Nations. https://peacekeeping.un.org/sites/default/files/peacekeeping/en/operationslist.pdf. Accessed 22 Jul 2023.

Our History UN Peacekeeping. n.d. United Nations. https://peacekeeping.un.org/en/our-history. Accessed 22 Jul 2023.

Peacekeeping Operations Fact Sheet. 2023. United Nations. https://peacekeeping.un.org/sites/default/files/peacekeeping_fact_sheet_may_2023.pdf. Accessed 22 Jul 2023.

Ten Rules: Code of Personal Conduct for Blue Helmets. 1995. United Nations. https://police.un.org/sites/default/files/ten_rules.pdf. Accessed 22 Jul 2023.

United Nations Department of Peacekeeping Operations. 1995. *Medical support manual for United Nations peacekeeping operation*. New York.

———. 1999. *Medical support manual for United Nations peacekeeping operation*. 2nd ed. New York.

United Nations Department of Peacekeeping Operations and Department of Field Support. 2015. *Medical support manual for United Nations field missions*. 3rd ed. New York.

United Nations Peacekeeping Operations Principles and Guidelines. 2008. United Nations. https://peacekeeping.un.org/sites/default/files/peacekeeping/en/capstone_eng.pdf. Accessed 22 Jul 2023.

Yu, Li, and Qiu. 2018. Overcoming new challenges in medical support for UN peacekeeping operations. *International Review of the Armed Forces Medical Services* 91: 20–28.

Chapter 9
Dual Professional Loyalty and Medical Ethics Outside Armed Conflict: The Physician-Soldier Dilemma

Innocent Okafor Eze and Uchenna Clara Innoeze

Abstract The foundational challenge of dual professional loyalty can create tensions for ethical medical practice. The apparent ultimate goal, exclusivism between medical and the military profession, creates conflicts of dual loyalty in the physiciansoldier that could impact negatively both the physician and the patient. The physician-soldier under the strict military command and control with the ultimate goal of dishing lethal harm on the enemies of the state for political objectives, ought solemnly to 'do no harm' and uphold other ethical demands (autonomy, justice and confidentiality) of medical profession which occasionally clashes with the military objectives. Furthermore, the physician part of the physician-soldier is regulated by a non-military authority which can sanction him without consideration to his peculiar situation if he breaches the ethics of the practice, hence placing him in a dilemma. In this paper, the authors tried to situate the intricate relationship between morality, ethics and the law, making a background for the topic hinged on the conflicts of medical ethics with the military demands in the physician-soldier, especially in garrison practice, which seems to have been neglected in preference to such debates during armed conflicts. The differing concepts of loyalty to superiors and organization in the military, and loyalty to principles and patients in the medical profession is highlighted. Drawing largely from years of experience as a physician-soldier in practice the authors point to the conflicts that threaten the ethical demands of patient autonomy, abuse of justice and neutrality, scientific validity, freedom of conscience and professionalism. Possible solutions include more education of the top military superiors through focused workshops and conferences and careful consideration of a "rankless" status for very ethical and more consciencebound profes-

I. O. Eze (✉)
Department of Obstetrics and Gynaecology, Nigerian Navy Reference Hospital, Ojo, Lagos, Nigeria

U. C. Innoeze
University of Nigeria Teaching Hospital, Enugu, Nigeria

Renaissance University, Enugu, Nigeria

D. Messelken, A. E. Barbar (eds.), *Challenging Medical Neutrality*, Military and Humanitarian Health Ethics, https://doi.org/10.1007/978-3-031-69398-4_9

sionals in the military. The erroneous but strongly held ideas by some military superiors that the physician part of the physician-soldier should be subsumed in the military as a mere tool at their disposal for achieving military goals, in total disregard to the ethics and humanitarian laws, should be more seriously and universally condemned.

1 Introduction

Ethical problems hinges on principle of morality, code of conduct within a society, standard of right and wrong. While morality is concerned with private and personal interactions, ethics regulates professional interactions. Law governs society as a whole, often dealing with interactions between heterogeneous members of the society. The law largely draws from the prevailing ethics in the society which crystalizes from the individual dynamic moral values largely informed by her ever-changing experience in the environment. While moral values are largely personal and sanctions are left to the individual conscience, breech of professional ethical codes may attract various degree of sanctions subject to approval by the law. The penalty for breaking the law when established in finality by court of competent jurisdiction usually ranges from mild to capital punishment according to the law, regardless of individual morals and ethical principles. Yet ethics, the 'conscience' and the moral 'police' of the society is "superior" to the individual morals, culture and the laws of a people as it appears universal in principles although relative in applications.

Ethical issues are 'eternal' philosophical problems that remain as important as unresolved. There is hardly any objective universally accepted standard of morality as humanity evolves, and yet human beings should have a meeting point of reference in relating ethically with one another to have a meaningful existence. Because principles of ethics are naturally and fairly general, they need to be interpreted and applied to specific situations. One role of ethics is to show how general principles apply in specific situations. However, some tension will always exist between universal principles and their application in particular contexts.

2 The Military and Medical Profession: Mutually Exclusive?

Medical and military professions are important and much respected careers worldwide. However both professions, simply understood, could be described (in principles) as fundamentally mutually exclusive professions. The primordial ethical principle for healthcare professions which starts with the dictum 'first do no harm' contrasts sharply with the military profession that has the ultimate function of applying lethal force against a nation's enemies to achieve political objectives (Bricknell and Story 2022). The military health professionals have rights and duties that are unique to the medical profession and are also subject to military law throughout their service in the armed forces. They have responsibilities to both their patients and their employer, and there could be a conflict between following military orders and medical professional codes of practice. Although employed in the armed forces, military healthcare professionals will be accountable to their national healthcare professional regulatory bodies for their ethical practice. Under the military command and control, the military medical professional is responsible for her ethical practices, without excuses. Disciplinary measures could be taken against her by the medical professional regulatory body strictly according to the guideline without consideration to the peculiarities of the military. Such measures may include withdrawal of practicing license (in extreme cases) which renders her useless to the military as a medical professional. The discussion and academic debate on dual professionalism and superiority conflict/complex encountered by the military medical professional outside armed conflict (in garrison practice) seems to have been relegated in favor of such conflicts in combat scenarios. The extent to which loyalty may be deflected from a patient to a third party (the military command) is greatly underestimated and has not attracted significant scholarly analysis (Benatar and Upshur 2008).

This paper is an effort to discuss and share experience as a military health professional on how military influence can impact on ethical medical practice and the "principle of neutrality" in military health facilities outside armed conflict. The authors hope sharing the experience will stimulate further debate on the subject towards resolving the conflict of dual professional loyalty in the best interest of the patient.

3 Background

> "We believe the role of the physician–soldier to be an inherent moral impossibility because the military physician, in an environment of military control, is faced with the difficult problems of mixed agency......" (Sidel and Levy 2003)

The physician-soldier under the strict military command and control with the ultimate goal of dishing lethal harm on the enemies of the state for political objectives, ought solemnly to 'do no harm' and uphold other ethical demands of medical profession which occasionally clashes with the military objectives. He wants to do the "right thing" but how does he know that which is right and how does he reconcile the conflict. The foundational challenges of apparently contrasting mandate and dual professional loyalty can create tensions for ethical medical practice by military medical personnel. Additionally, the combination of the 'two strange bed-fellows' in one person could result in crises of value and cause significant mental distress to the military health professional. This could compromise care in the best interest of the patient.

Medical Ethics are the principles or norms that regulate the conduct of the relationships between medical practitioners and other groups with whom they come in contact during the course of their practice (Adimma 2016). Military Medical Ethics

(MME) could be defined as an ethical and moral effort towards resolving the dynamics and delicate interplay between military demands and ethical medical practice by military health professionals both within and outside armed conflict.

While much emphasis on the academic debate for MME is placed on compliance with International Humanitarian Law (IHL) especially during armed conflict, the non-operational component of MME must not be neglected. This is because the majority of a military healthcare professional's clinical career is likely to be spent on military health facilities outside armed conflicts zones (Marshal and Miron 2021). Here, the military medical professionals, similar to colleagues in armed conflict environment, do often encounter situations in the military health facilities where, occasionally, clinical decisions are taken more to please military superiors contrary to ethics and evidence-based practice. Furthermore, there is need to highlight and address the usually 'silent' but important areas of conflicts in upholding ethical medical practice by military medical professionals in military health facilities outside armed conflict. The areas of conflicts include, but not limited to ethical demand for autonomy, justice, and evidence-based practice. Also, the issues of 'discordance' of command/hierarchy and dual professional leadership crisis among military medical professional in garrison practice should be addressed. The consensus has long been that despite the different contexts, the ethical responsibilities of health care professionals to their patients are the same in the military context as in civilian life. Living up to such expectations is more challenging when health care decisions can be influenced or directed by the requirements of military, but medical professionals under the military command are urged to rise above these pressures and to honor their ethical responsibilities even if this jeopardizes them personally (Benatar and Upshur 2008).

4 Loyalty in Military and Medical Profession

Synonyms for loyalty include: allegiance, devotion, fidelity, faithfulness, etc. Loyalty implies faithfulness that is steadfast in face of any temptation to renounce, desert, or betray. (Collins English online-dictionary). Loyalty in itself is not traditionally considered a virtue by the ancient philosophers (Olsthoorn 2019). This can easily be understood when "loyalty" is placed side by side with traditional virtues like justice, honesty, charity, prudence, fortitude, temperance which stand on their own as virtue, more or less. While the 'virtue' of loyalty is relative to the object, be it group, person, or principle, the traditional virtues are almost absolute than relative. For instance there is group loyalty among criminal gangs and one could also be loyal to one's positive principles. Whereas in loyalty the end justifies the value, the traditional virtues are essentially universal, absolute and do not require contextual interpretations.

Loyalty is about the most valued 'trait' in the military. Indeed, 99.9% loyalty in the military is not a pass mark; it must be absolute or claimed to be so for every military personnel to be in good standing with the organization. Indeed, any military personnel whose loyalty to superiors is in doubt is not likely to make good progress in the job. Here, loyalty is to the organization, superiors, the state and the 'people'. For the medical profession, loyalty is to the principles and ethics of the profession largely in the interest of the third party (the patient). Here, one can talk about loyalty to upholding patients' right to autonomy, justice, respect for human life etc. above the interest of colleagues, superiors, group, organization etc.

Armed forces socialize their employees thoroughly into the organization, which contributes to the strong loyalty military personnel feel towards each other and the organization. That military personnel is predominantly trained in house, whereas doctors and other professionals receive most of their formal professional training before entering their job, makes this socialization into the organization, instead of into a profession, easier. As a consequence, different militaries have different organizational values (often still service specific), but there are no values of the military profession as such (Olsthoorn 2019).

In the modern world, however, health professionals are frequently placed in settings where they are asked to weigh their devotion to patients against service to the objectives of government or other third party (London et al. 2006) The moral dilemmas military medical personnel face stem from a conflict between loyalty to one's colleagues or organization on the one hand and a more universal ethics, one's own conscience and moral standards on the other. Faced with a choice between loyalty to a group and loyalty to a profession, it lies at hand that the latter is the one to act upon, since it has a wider scope and includes more than just the interests of colleagues or an organization. In many aspects, the professional loyalty that characterizes medical ethics has a lot more in common with loyalty to principle than with the group loyalty that militaries further (Olsthoorn et al. 2013).

5 The Conflicts of the Military with Medical Ethics

Now, we look at the various areas of conflicts for the military medical professionals in the light of the pillars of biomedical ethics amplified in the World Medical Association (WMA) international code of medical ethics 2022.

5.1 *'First, do no harm!'*

> The physician must never participate in or facilitate acts of torture, or other cruel, inhuman, or degrading practices and punishment (WMA Oct. 2022).

The ethical dictum, "*Primum non nocere*" (Latin) is largely understood and primarily interpreted that physicians should not use their knowledge to do harm. This clearly forbids the doctor from the use of drugs, clinical knowledge or any other means to intentionally harm his patients or any other (even from time of conception depending on the national or state abortion law). It also implies that, in medical practice, it is better to do nothing than do something that may harm or worsen the patient's health condition.

In times of armed conflicts, military medical professional alongside other humanitarian players are given special protection under the International Humanitarian Laws as "neutral" harmless participants in the conflict. They are non-combatant military personnel that should ideally give medical assistance to the wounded soldiers and civilians who may need some medical attention from both sides of the conflicts. Even when they bear light weapons as allowed by IHL, it is meant for protection of self and patients should the need arise. The physician-soldier loses such protection under the law if he steps outside these set boundaries to take sides and do harm. The practicability of the ideals of the paradox of a "noncombatant-harmless soldier" neutral to the conflict goals on both sides and yet loyal to a state and her armed forces in the conflict has been questioned and debated. Some argue that such ideals are utopian and "delusional" as the conflicts of dual loyalty in the "physician-soldier" is often naturally resolved in favor of the state and goals of the armed forces than the medical ethics. However, the details of the conflicts of dual professional loyalty and medical ethics in combat zone has been addressed by other authors and not within the scope of this paper which is rather focused on such conflicts in garrison practice.

The medical ethical principle of "do no harm" could rightly be stretched to forbid the physician from participating in any form of "violence" *(torture, or other cruel, inhuman, or degrading practices and punishment)* against her patients or any other as much as possible. If this is right, then the following example may not be too far out of context.

As officer cadets during the military training, the military medical officer in the training college actively participated in "training" the cadets which involved activities/treatments that a civil mind could describe as "*torture, cruel, inhuman and degrading*". Before giving them medical attention, the cadets were subjected to such treatments by the military medical officer even when they visited the medical center as patients. Although it could be considered part of training, and judged right within military ethics, it was a rude shock how a medical doctor could participate in such "cruel treatment" to his patients in the name of performing military duties of training the cadets. From my assessment then, the military doctor often looked unhappy. He was a 'victim' of the conflicts of dual professional loyalty. As a medical military professional in military training institution his primary task was to attend to the health needs of the cadets. But in doing this, he is often caught in the web of military duties of "training" the cadets in the form that may be perceived as "torture" even in the clinic setting as they seek medical attention. It was a conflict of duty in principles. A care-giver that "harms" the one he cares for. What a contradiction! It was no surprise that most of the "patient cadets" hardly trusted him as a doctor and preferred seeing civilian doctors (if they had the choice) when they fell sick. This breach of trust can indeed compromise care in the interest of the patient.

Should a military medical professional participate in such "training" of his patients even as cadets? Reflection on this question may point to ideas that may regulate the activities of the physician-soldier that finds himself in such a professional conflict of duty as above. In the opinion of the authors, (supported by the guideline of the International Dual Loyalty Working Group, 2002) the physician-soldier in military training institution should focus on his foremost and overriding duty of attending to the health needs of the cadets, essentially in line with the ethical prescriptions of the physician-patient relationship. However, this should be without prejudice to the permitted situational ethical consideration of the "greater good", which may allow some relative modification in application especially in areas of patient autonomy and confidentiality. Nevertheless, he should leave the "training" of cadets (especially at the time they present as his patients) to other soldiers free from the burden of dual professional conflict of duty.

5.2 *Autonomy*

"*In providing medical care, the physician must respect the dignity, autonomy, and rights of the patient. The physician must respect the patient's right to freely accept or refuse care in keeping with the patient's values and preferences. The physician must respect the patient's decision to withhold or withdraw consent at any time and for any reason*". (WMA Oct 2022).

5.2.1 Patient's Autonomy

The autonomy of capable persons is central and a very vital pillar in medical ethics. No form of care, no matter the perceived benefits, should be imposed on a capable patient**.** The basis of this principle is connected to the idea that all persons possess intrinsic worth, independent of any special circumstances that confer value. Actively showing respect for the autonomy of others involves due appreciation of their capacities and perspectives, including their rights to hold views, to make choices and take actions based on personal values and beliefs. This is also conceptualized as Patient Preferences (PP). In a military setting where order and obedience to superiors are paramount, autonomy of the patient (the military personnel) could easily be overlooked and compromised.

A complex dual loyalty problem may arise in jurisdictions where military service is voluntary and members of the armed forces are generally held to have voluntarily waived some of their rights by choosing to join the armed services. As patients, they take some responsibility in advance for deciding the extent to which they are willing to "give up" their rights, including, for example, the right to doctor-patient confidentiality. ***However, this agreement does not mean all of a military patient's rights are necessarily waived and health care providers should therefore not exceed what is "necessary" in any disclosure. Indeed, it is debatable whether the waiving of rights by conscripts could be reasonably accepted as voluntary*** (Dual Loyalty Working Group 2002).

The military organizations are said to be justifiably, to a certain extent, paternalistic organizations (Wolfendale and Clarke 2008). This may be excused on the unique ground of special duty of care the military organization possesses, with respect to its members. Specifically, the Navy operates a Divisional System where an officer should take far reaching responsibility for military personnel under her. This is a positive ideology meant to care and support the military subordinate. But when such 'care' does not respect the views and preferences of the subordinate, despite the 'good' intentions of the superior, it could be perceived as a form of Military Paternalism.

Paternalism is the policy or practice on the part of people in authority of restricting the freedom and responsibilities of those subordinate to or otherwise dependent on them on their supposed interest. (Online oxford dictionary). It is the interference with the liberty or autonomy of another person, with the intent of promoting good or preventing harm to the person (examples: laws on wearing seat belts, helmets and "harmful" drug use).

5.2.2 Medical Paternalism: Where Can It Be Justified?

Medical Paternalism could simply be defined as overriding PP in the perceived best interest of the patient. It has been argued that while paternalism may be justifiable in Tactical Combat Care ("under-fire care") and possibly in Advanced Trauma Management (ATM) (likely within the operational area where there may not be the 'luxury' of considering PP), this may not likely be the case with Outside Combat Care that provides similar working conditions as in civil health institution (Nathanie Sharadin 2022). There are ethical principles in addressing "under-fire care" that may "violate" patient autonomy and preferences. Outside combat care may also present scenarios similar to civil health institutions where patient autonomy may be overridden, like in cases of a minor or "incapable" persons. Here, ethics and the law provides a leeway through the attorney or the surrogate decision maker who takes a decision on behalf of the minor or incapable patient in "good and objective conscience".

Under pressures of military command structures, medical personnel face significant dual loyalty conflicts when performing evaluations for legal or administrative purposes that have serious implications for victims' human rights (Leslie London et al. 2006). In military health institutions, a patient's perception of freedom of choice over medical treatment might be constrained by rank or other power differentials if the healthcare provider outranks the patient or vice versa (Coleman 2020). This may come from a superior military medical professional (intentionally or not) usurping the junior military personnel right of autonomy as patient. For example, the decision to admit and administer prescribed treatment to a military subordinate (patient) by superior military medical professional is more or less a standing order to which the patient tactically has no choice of refusal. It could also result from the patient herself, who out of ignorance of such right or more often for fear of victimization shy away from exercising her right of autonomy. A non—medical military superior could 'order' a subordinate to undergo a medical check/test or be subjected to certain medical treatment against a valid voluntary informed consent of the subordinate. This often could be for administrative interest or the military law/instructions/order to which both the medical military personnel and or the patient (military personnel) are subject. Restrictions on the choice of healthcare provider due to military rules or payment systems that prevent access to care outsides the military health system may exacerbate this situation. Military personnel should not be ordered to receive a specific medical treatment, although they can be ordered to attend for medical care (Marshal and Miron 2021).

5.2.3 The Autonomy of the Physician Under Military Command

> How can military medical professionals most appropriately balance their obligations to their patients against their obligations as military officers to help commanders maintain military readiness? (b) How much latitude should military medical professionals be given to refuse participation in medical procedures or request excusal from military operations with which they have ethical reservations or disagreement? (United State of America Defense Health Board 2015)

How 'free' indeed, is the military medical professional under military command in garrison practice? Can the military doctor freely select and manage his patients based on competence, capacity and fair clinical judgment without undue interference by the military command? Can he freely refer cases he can't manage in good conscience, devoid of the interests of the military superior? Are the clinical duties of military medical doctor under military command considered strictly as "military duties"? Could he be charged for refusal to perform "military duties" under the armed forces law, if the doctor refuses to carry out such clinical duties he considers unethical and or unprofessional under military command? Where do we draw the line? Finding 'universal' answers to above questions may be challenging in application in different contexts. But we need to proffer guidelines in addressing them as they are valid practical challenges the military medical professionals under military command do encounter both within and outside armed conflicts.

The third party interests, no doubt, do often come between the ethical relationship between the physician and the patient. 'Who pays the piper dictates the tune' they say. The physician, as an employee both in the military and civil health institutions, may be under pressure to adhere to institutional goals which may compromise the ethical relationship between the physician and the patient. The integrity of the physician and his loyalty to ethical medical practice is usually on trial when faced with choice of protecting the third party interest to keep his job and ethical mandate of loyalty to his patients. In the military setting this conflict of dual loyalty is even more intense and difficult considering the highly regimented, command and control, and restrictive nature of the job. Indeed, a physician-soldier in garrison practice needs extra courage than the civilian counterpart to uphold loyalty to patients and ethical practice against the goal or the opinion of the military commander. In the opinion of the authors, the International Committee on Military Medicine (ICMM) and order relevant bodies should consider focus on more education of the top military commanders on the need to provide and promote framework that will encourage the physician-soldier uphold ethical practice without fear of victimization. They or their representative could be invited for workshops to address such issues. This is because some superior military officers, lacking in the basic knowledge of medical ethics and its overriding nature over military command and control, uphold that medical professional part of the physician-soldier should be subsumed in the military as a mere tool at their disposal for achieving military goals in total disregard to the ethical guidelines of the medical profession.

5.2.4 Medical Hierarchy and Military Ranking: The Leadership Conflict in Clinical Setting

Another point of conflict amongst the military medical professionals that needs attentions is that of the discordance in medical hierarchy and military command. Medical profession, like the military, acknowledges and respects hierarchy and chain of command. Among the doctors, for instance, the house officer reports to the senior house officer who reports to the junior resident doctor who is under the senior registrar who reports to the medical consultant who may then report/discuss to/with more senior medical consultants. In the military, the chain of command all things being equal, is strictly according to seniority in ranks. But what happens when there is “discordance” of hierarchy in a “dual hatted” military medical professional? For example, a senior ranking military personnel with lower medical qualification working in the same medical unit with a junior ranking officer with a superior medical qualification. Who takes charge? Who reports to who in clinical matters? Depending on the environment, maturity, individual professionalism and tact, the answer may be simple or difficult just as the working relationship. In a military setting that ‘overemphasizes’ military rank over and above professionalism, conflict/crises of leadership may arise. The superior ranking military officer may resist professional medical instructions from the military junior, but not from a superior medical professional. Such conflicts, sadly, can rub off negatively on ethical medical practice and patient care.

Some tertiary military health institutions are equally saddled with additional responsibility of training health professionals including high ranking military medical officers. The officers could be at various lower cadre in the medical profession including house officers/interns and resident doctors who want to pursue a medical specialist training as serving military personnel in military health institution. When the trainers (with higher medical qualifications) are junior to the trainees in military hierarchy it provides a difficult platform for efficient training. The confusion/conflict of chain of command and leadership comes to play. If this is not properly resolved, it undermines good training and product efficiency with the patient ultimately bearing the consequences.

There is need for the military medical ethics to provide guidelines that will address such conflicts in the best interest of ethical patient care. The foremost guideline by the (International Dual Loyalty Working Group, 2002) which state that: "*The military health professional's first and overruling identity and priority is that of a health professional*" could be further detailed to provide a framework that could address such a scenario. The authors propose that the military medical professionals under training in military health institutions should be "rankless" while in training. This will mitigate the tension of the conflict/confusion of military/medical command/hierarchy discordance. Also, healthcare team leadership and clinical decisions in military health facilities should be based on certified medical professional competence and experience and not only on military ranks.

5.3 Justice (Impartiality)

> *The physician must practice medicine fairly and justly and provide care based on the patient's health needs without bias or engaging in discriminatory conduct on the basis of age, disease or disability, creed, ethnic origin, gender, nationality, political affiliation, race, culture, sexual orientation, social standing, (rank)* or any other factor (WMA* 2022).

The ethical demand for "justice" in medical practice, in the opinion of the authors, is the most abused pillar of Clinical Ethics both in civil and military setting. Health professionals at all levels and institutions brazenly breach this ethical code of conduct. Some believe the concept of 'absolute justice and fairness' and indeed the "principle of neutrality" in medical practice (and in fact the larger human society) is more utopian, delusional, ideal, and theoretical than practicable.

In modern medical practice, where ability to pay largely determines the care (timing, quality…etc.) the patient gets, the principle of justice is nipped in the bud. It remains a pipe dream for the poor. Most low income countries do not have efficient Health Insurance System (HIS) which should mitigate the burden of health care bills on individuals. The patients pay out of pocket. The largely poor population who cannot afford the bill is denied the care. The poor, shamefully, die from very preventable diseases. In richer and more advanced society with HIS in place, the system only covers the basic care while high cost health bills are borne by individual. Those who can't afford it won't access it!

So where is the "justice" in the healthcare? The answer may be subject to philosophical and ethical debates which is beyond the scope of the discussion. So I narrow down to my topic which borders on how dual professional loyalty in the military setting impacts on the ethics of justice and principle of neutrality outside armed conflict.

The abuse of justice and neutrality in medical practice is better envisaged in military setting. To be fair, HIS is in place for all military personnel despite the rank and type of commission but one finds subtle discrimination in application. Ideally, care should be offered based on "patient's need only, and not any other considerations". This may imply that patients with similar health needs should be attended to on the basis of "first come first served". In practice this is a "fallacy." The military personnel patient is given priority over the civilian, all things being equal. Could this be justified? For the military personnel, the officers and the other rank are not given a fair and just attention based on health needs only. The officers are often the priority, all things being equal. Among the officers and the other ranks, with similar health needs, priority of care is often determined by ranks and command. Spouses and relatives of superior military personnel are also given priority over that of the inferior personnel and civilians with similar health needs. Indeed, the timing and quality of healthcare is largely determined by power differentials, be it military, political or economic.

The military medical personnel, like colleagues in the civil institutions, face the conflict of the dilemma of upholding the ethical practice or 'stepping on the toes of the powerful and the authority with the consequences.

According to Olsthoorn (2019), "*Group loyalty is not a value, and the choice between loyalty to colleagues and acting impartial is therefore not a moral dilemma, but a test of integrity: it is clear what the right thing to do is, but there is considerable pressure to choose the wrong course of action*".

5.4 *Scientific Validity*

Scientific validity ensures professional competence and scientific soundness in medical research and practice. It entails physician should uphold evidence-based practice.

> Physicians must indicate if their own opinions are contrary to evidence-based scientific information (WMA 2022).

5.4.1 "A Clinical Decision" Influenced by a Military Superior

Physicians must take responsibility for their individual medical decisions and must not alter their sound professional medical judgments on the basis of instructions contrary to medical considerations. (WMA 2022)

It sounds, indeed, odd that a clinical decision by a competent medical professional in the military could be influenced (negatively by the way) by a military superior. But the physician-soldier practices his medical profession in an environment where he has to constantly weigh his options: loyalty to ethics and patient or loyalty to superior, and organization. The power differentials may likely often favor the later. A colleague (military medical professional) recently shared his experience of how a superior non-medical officer (patient) 'ordered' him to prescribe some antihypertensive of his (patient's) choice. The superior officer was not his patient and he had no knowledge of his medical history. The doctor initially resisted the order as unprofessional, but eventually succumbed to pressure for fear of victimization. The superior officer was very influential and the military medical doctor was not ready for his trouble. I wondered how a medical doctor could be ordered to append his signature to a prescription, in total disregard to evidence-based clinical methods. But then I was no better having occasionally mitigated my clinical judgment by the need to please a military superior.

In a military hospital where I practiced as a consultant obstetrician, I was occasionally under pressure by a medical military superior to offer caesarean section (CS) to parturient against evidence-based clinical judgment. The medical military superior (though not an obstetrician) believed CS was "safer" for both mother and fetus than labor, which is dynamic and could result in complications. She believed that was necessary to forestall maternal and fetal morbidity and mortality which could bear negatively on her command. But evidence-based practice demands that there should be valid indication for CS with valid informed and voluntary consent of the patient. I must choose between the ethical evidence-based practice and 'unethical' easy recourse to CS to please the military superior and be seen as a 'loyal' subordinate. The dual loyalty conflict was often resolved in favor of military than in the interest of evidence-based practice. The implication was very high CS rate (> 50%), with associated increased cost to the patient, longer hospital stay, increase in repeat CS, limitation on preferred family size, and unnecessary exposure of the parturient to the potential short and long term complications of CS. The ethical necessity of scientific validity is breached but that is not all. Indeed to 'convince' the parturient she needed the CS the doctor may not be objective in counselling her to effect ethical informed voluntary consent. This, again, breaches the ethical necessity of "veracity and fidelity" ("tell all the truth and do what is promised").

There are other occasions of such compromise where the physician- soldier may be intimidated under the military command to mitigate clinical report and even laboratory investigations to suite the goal of a military superior against the ethics. For instance, during military recruitment or enlistment as the case may be, medical screening for candidates is a very serious exercise and indeed a very limiting point of entry for candidates who wish to join the force. The physical and mental rigors of military training demand that only the medically fit can safely embark on such training. There are pre-set standards and guidelines for medical fitness test and selection criteria. However, very powerful people, including superior military personnel, have their preferred candidates who must make it at all costs. The healthcare worker under the military command, entrusted with the responsibility of this exercise, is occasionally under pressure to compromise set-standard or true fitness outcome to favor the candidates of powerful superiors. This is unjust and unethical to say the least, but a more serious disservice to humanity and even the 'favored' but ignorant candidate. The implications of such maneuver include but not limited to higher but avoidable morbidities and mortalities during training, increased medical bills to the Service and unhealthy force. Practicable policies should be in place to encourage the physician-soldier in such an environment to uphold ethical practice with minimal consequences. We conclude by outlining the guidelines for the military on dual loyalty as packaged by International Dual Loyalty Working Group, 2002.

6 Guidelines for the Military on Dual Loyalty

1. The military health professional's first and overruling identity and priority is that of a health professional.
2. Civilian medical ethics apply to military health professionals as they do to civilian practitioners.
3. The military health professional should adhere to the principle of confidentiality in a manner consistent with practice in civil society.
4. The military health professional is a member of the national and international health professionals' community.
5. The military health professional should treat the sick and wounded according to the rules of medical needs and triage.
6. Health professionals should not participate in research or development of chemical or biological weapons (CBW) that could be used for purposes of killing, disabling, torturing, or in any way harming human life.
7. The military health professional should refrain from direct, indirect, and administrative forms of cooperation in torture and cruel, inhuman and, degrading treatment and punishment at all times, including in wartime and during interrogation of prisoners.
8. The military health professional should refrain from direct, indirect, preparatory, and administrative participation in capital punishment, both within the military court martial system and elsewhere.
9. Military health professionals should report violations of human rights that interfere with their ability to comply with their duty of loyalty to patients to appropriate authorities and report human rights violations perpetrated by their own troops as well as by others.
10. The health professional should not engage or participate in any form of human experimentation among members of military services unless the research will provide significant health and other benefits for military personnel and facilitate promotion of their human rights.
    (Source: International Dual Loyalty Working Group 2002)

The dual loyalty guidelines hold that medical ethics during wartime are not fundamentally different from those applicable in peace, and that the processes and mechanisms required to determine permitted deviations from the norm are not particular to the military context. Rather than generating a new paradigm for ethical practice, which is based on the creation of what is essentially a non-practice model of undivided loyalty to a military commander, resolving the ethical dilemmas of dual loyalty during armed conflict (and outside armed conflict) is better served by an approach grounded in human rights, which ensures that the duties to respect and protect human dignity remain at the core of health professional practice.* (Leslie London et al. 2006)

7 Conclusion

In this paper, the authors have tried to situate the intricate relationship between morality, ethics and the law, making a background for the topic hinged on the conflicts of medical ethics with the military demands in the physician-soldier especially in garrison practice. The apparent professional exclusivity between medicine and the military usually create conflicts of dual loyalty that impacts negatively on both the physician and the patient. The physician-soldier under the strict military command and control, with the ultimate goal of dishing lethal harm on the enemies of the state for political objectives, ought solemnly to 'do no harm' and uphold other ethical demands of medical profession which occasionally clashes with the military objectives. But then, the physician in the soldier is regulated by a non-military authority which can sanction him without consideration to his peculiar situation if he breaches the ethics of the practice, hence placing him in a dilemma. The differing concepts of loyalty to superiors and organization in the military, and loyalty to principles and patients in the medical profession was highlighted. Autonomy of capable patients an ethical necessity in medicine is usually flaunted in the garrison practice. Drawing largely from the experience as a physician-soldier in practice, authors pointed to such areas of conflict as "ordering" or "intimidating" capable patients, especially the military personnel, to undertake medical investigations and treatment against their valid informed consent, at times to serve military objectives. Another pertinent issue is the freedom of the physician soldier to take clinical decision based on credible evidence and international best clinical practice or at his discretion in good conscience without influence and interest of the military superior. The authors posed an important question: "Are the clinical duties of the physician-soldier strictly under "military duties" for which he can be charged according the military law at the discretion of the military superior"? A very important aspect of the dual loyalty conflicts is in the discordance in medical hierarchy and military command. Medical superiors of inferior ranks are subject to the superior military personnel with inferior medical qualification in the same unit at times even in clinical matters. The military rank is emphasized over and above medical qualification which occasionally compromise ethical medical and best clinical practice. The issue of justice (impartiality/neutrality), an important pillar of the 'quadripod' upon which the bioethical principle stands, in the opinion of the authors, is the most violated ethical principle both in the civil and military environment. Money and status seems to determine to a large extent who gets the best of attention and care in total disregard to this ethical guide. The authors finally proffer solutions believed to help the physician-soldier survive his ethical practice in the military environment. Such possible solutions include more education of the top military superiors through focused workshops and conferences. A careful consideration of a "rankles" status for very ethical and more conscience-bound professionals in the military like medical doctors and religious leaders could be a bold step. The erroneous but strongly held ideas, by some uninformed military superiors, that the physician part of the physician-soldier should be subsumed in the military as a mere tool at their disposal for achieving military goals, in total disregard to the ethics and humanitarian laws, should be more seriously and universally condemned. Global humanitarian policies to address the conflicts of dual professional loyalty among the military medical personnel will impact heavily on resolution of the conflicts in the best interest of the patient and humanity.

Recommendations

1. More practicable policies should be developed by relevant global agencies to address the conflicts of dual professional loyalty among the military medical personnel that impacts negatively on ethical medical practice.
2. There should be more education of the top military personnel on military medical ethics through workshops and conferences.

References

Adimma, J.I.B. 2016. *Bioethical issues related to reproductive health and sexual right.* Lecture delivered at revision course in obstetrics and gynaecology for part 1&2 of the West African College of Surgeons, Abuja Nigeria.

Benatar, S.R., and R.E.G. Upshur. 2008. Dual loyalty of physicians in the military and in civilian life. *American Journal of Public Health* 98 (12).

Bricknell, M., and R. Story 2022. An overview to military medical ethics. *Journal of Military and Veterans Health 30 (2): 7.*

Coleman, N. 2020. The impact of the duty to obey orders in relation to medical care in the military. In *Ethics of medical innovation, experimentation, and enhancement in military and humanitarian contexts*, 37–52. Springer.

London, L., L.S. Rubenstein, L. Baldwin-Ragaven, and A. Van Es. 2006. Dual loyalty among military health professionals: Human rights and ethics in times of armed conflict. *Cambridge Quarterly of Healthcare Ethics* 15 (4): 381–391.

Marshall, M.C., and M. Miron. 2021. Medical ethics for the military profession. *Revista Científica General José María Córdova* 19 (36): 851–866.

Olsthoorn, P. 2019. Dual loyalty in military medical ethics: A moral dilemma or a test of integrity? *Journal of the Royal Army Medical Corps* 165 (4): 282–283.

Olsthoorn, P., M. Bollen, and R. Beeres. 2013. Dual loyalties in military medical care – Between ethics and effectiveness. In *Moral responsibility & military effectiveness*, ed. H. Amersfoort, R. Moelker, J. Soeters, and D. Verweij, 79–96. Leiden: T.M.C. Asser Press.

Project of the International Dual Loyalty Working Group. 2002. *Dual loyalty in health and professional practice: Proposed guidelines and institutional mechanisms. Physicians for human rights; and Cape Town, South Africa.* University of Cape Town Health Sciences Faculty.

Sharadin, N. 2022. Personalized patient preference predictors are neither technically feasible nor ethically desirable. *The American Journal of Bioethics 24 (7): 62–65.*

Sidel, V.W., and B.S. Levy. 2003. Physician-soldier: A moral dilemma. In *Military medical ethics*, ed. T.E. Beam and L.R. Spracino, vol. 1, 293–312. Falls Church: Office of the Surgeon General. at p. 296.

Wolfendale, J., and S. Clarke 2008. Paternalism, consent, and the use of experimental drugs in the military. *Journal of Medicine and Philosophy 33 (4): 337–355.*

World Medical Association (WMA). 2022. *International code of medical ethics.* Berlin: Seventy third WMA General Assembly.

Chapter 10
Providing Medical Care to Further Non-medical Ends

Philisiwe Precious Ncayiyana

Abstract Providing medical care to further non-medical ends involves the use of medical expertise to achieve goals that are not strictly related to promoting the health and well-being of patients. This practice raises a range of ethical concerns regarding the purpose and scope of medical care, the relationship between health and national security, and the obligations and duties of medical professionals to their patients, professional bodies and society at large. Special Forces types of operations in support of other military objectives bring about this dilemma. The well documented case of then Lieutenant Colonel (Dr) Wouter Basson, branded "Dr Death" by the media, is a classic example of this dual loyalty dilemma. The study found that the ethical issues of whether military doctors can dissociate their medical and military personas has important implications. The medical council's provisional guilty findings against Dr. Basson seem to indicate that those elements cannot be dissociated. This decision will have a profound impact on the future of Special Forces doctors after they leave the operational environment. Application of ethics in military medical practise is complex and controversial and this leaves the practitioners exposed to possible sanctions by their statutory bodies. Perhaps there should be different kinds of ethics for Military Medics. This is by no means promoting unethical conduct.

1 Introduction

Militaries across the world comprise of different units and services like the Army, Air Force, and Navy. These are major services that enable a country to secure its borders and partake in United Nation's or Military alliance operations. Vital to these military organization's success are a host of professional bodies or personnel that provide services according to mustering such as engineers, accountants, lawyers and healthcare practitioners. These professionals are accountable to military

P. P. Ncayiyana (✉)
South African Military Health Services, Midrand, South Africa

D. Messelken, A. E. Barbar (eds.), *Challenging Medical Neutrality*, Military and Humanitarian Health Ethics, https://doi.org/10.1007/978-3-031-69398-4_10

command as well as civilian statutory bodies. In South Africa, one of the healthcare practitioners' statutory bodies, is called the Healthcare Professionals Council of South Africa (HPCSA).

Health Care Practitioners (HCPs) in the military are sometimes deployed in non-peace keeping, and non-medical operations that bring out the dual mandate and often demand dual loyalty to both the military and professional medical bodies. The Special Forces' types of operations and activities in support of other military objectives bring about this dilemma and controversy.

Providing medical care to further non-medical ends involves the use of medical tools and expertise to achieve goals that are not strictly related to promoting the health and well-being of patients. This practice raises a range of ethical concerns regarding the purpose and scope of medical care, the relationship between health and national security, and the obligations and duties of medical professionals to their patients, professional bodies and society at large. One of the examples of providing medical care to further non-medical ends is the use of medical personnel and resources to support military operations. These actions may be seen as justifiable in the face of a greater good, but they present ethical questions about the roles and obligations of medical professionals in the military.

A world famous example of using medical care to further non-medical ends happened during Adolf Eichmann's capture by the Israeli secret services. Eichmann is considered the architect of the Holocaust. He was a Nazi official who helped Germany carry out the mass murder of six million Jews during World War II (Blakemore 2019). He escaped from Germany to Argentina after the war but was later tracked down by the Israeli secret services and captured back to Israel. Dr. Yonah Elian helped sedate Eichmann during this extraction mission. Israeli operatives then dressed Eichmann in uniform like an Israeli flight crew and were able to pass through airport security authorities into the plane and off to Israel successfully (Estrin 2019).

Dr. Yonah Elian was an Israeli anaesthesiologist, (Estrin 2019), a civilian doctor who was just recruited for this mission. How he sedated Eichmann was considered ingenious because it is said that Eichmann was semi-conscious, though sedated he could still respond to certain orders (Estrin 2019). Dr. Elian never spoke about Eichmann's operation ever until the very end when books and articles were published naming him as the doctor that sedated Eichmann. Dr. Elian even refused accepting awards in honour of his contribution to the success of the mission (Estrin 2019) and finally told his son that he had acted against the Hippocratic Oath, the doctors' pledge to do no harm.

1.1 *The Case of Dr. Basson*

This article will touch upon the well-documented case of Lieutenant Colonel Dr. Wouter Basson (later promoted to Brigadier General), who was branded "Dr Death" by the media and subsequently by certain individuals, communities and political activists in South Africa.

Dr. Basson joined the South African Defence Force (SADF) in 1975 as a medical doctor and qualified as a cardiologist in 1980 (Adeaga 2022). By the early 1980s, he had risen to the rank of lieutenant colonel, and it was then that the military asked him to head the SADF's top-secret chemical and biological weapons (CBW) program. He agreed and ran it for the next 12 years. According to one of the people interviewed, Dr. Basson's orders were to put up a defence chemical capability of which he did.

In the early 1980s it was noted that some Cuban soldiers were carrying Chemical and Biological (CB) protective personal equipment (PPE) and were suspected of using chemical weapons against The National Union for the Total Independence of Angola (UNITA). It was feared that such weapons would be used against SADF troops in the Operational Area. At that time the SADF had no PPE against CB weapons and due to international sanctions against the South African apartheid policy and government, SADF could not purchase such items in sufficient quantities on the open market (Blunden 2023). The Surgeon General then instructed Dr. Basson to put the necessary measures in place to safeguard SADF troops, including acquiring or manufacturing the needed items, which would obviously include sanction busting (Blunden 2023). As a result, Dr. Basson became the director of South Africa's defensive Chemical and Biological warfare programme, called Project Coast (PC) during the 1980s. This program was established in 1981 after the Biological and Weapons Convention (BWC) of 1972.

Dr. Basson was criminally charged almost 20 years after Project Coast was first established. His trial started in 1999, with 67 charges, including drug possession and trafficking, fraud, conspiracy to commit murder, murder, and theft (Jackson 2015). During the trial some very concerning allegations against him were published, including that the main objective of Project Coast was to annihilate the black population, and to developing a "Black bomb" that would selectively attack black people (Marbot 2021).

The trial lasted 30 months and in April of 2002 the Judge acquitted Dr. Basson and dismissed all the charges against him. The state appealed against the judgement but the Supreme Court of Appeal as well as the Constitutional Court refused to order a retrial (Swart 2008). That was 21 years ago. Since the South African Constituion includes the right not to be tried for a crime that a person was previously found guilty or acquitted of ("double jeopardy" clause) (South African Government 1997). Dr. Basson cannot be tried for these charges again.

There are those South Africans who are convinced that Dr. Basson should have been found guilty and insist that the 'Dr Death' title is still most appropriate (Njanji 2014). Others agree with court's findings and think that this title is inappropriate. However, in the light of the not guilty finding on all charges by a competent court, from a legal point of view, unless another court of law arrives at a different conclusion the "Dr Death" title should not be used as it could constitute libel which is damage to one's reputation.

The controversy around Dr. Basson started in 2002 and is yet to be concluded by the HPCSA. This indecision to proceed and conclude the case is unfair and unwarranted because a dark cloud of despise, suspicion and presumption of being unethical hangs over the practitioner. In 2021 a political party and a civil organization led the protest against a private hospital operator to prohibit the doctor from accessing their facilities (Hlati 2021).

The issues raised in Dr. Basson's case pose some challenges in balancing medical neutrality and impartiality. It has been a thorn in the eye in South Africa for the longest time. As part of his defence Dr. Basson told the court that he was following orders and the court could not find evidence that personally implicated him in manufacturing and distribution of biological weapons and toxins. The protracted nature of Dr. Basson's prosecution at the criminal courts and medical council highlights the difficulty of finding neutrality and harmonising dual loyalties for military health care practitioners.

This case evokes varied and often vehemently opposing opinion among health professional and political analysts and remains prominent in the modern history where medical means are considered to have been used for non-medical ends. Some members of the public, including 27 medical doctors, asked the HPCSA to take action against Dr. Basson and 40 other doctors were implicated in unethical conduct in the testimony before the Truth and Reconciliation Commission (TRC) (Baleta 2012). People's Health Movement South Africa described Dr. Basson's lack of remorse and his denial of wrong-doing that demonstrated lack of understanding of right or wrong (Njanji 2014).

1.2 The Crux of the Matter

The burning question addressed in this article is, should there be different kinds of ethical standards for military medical personnel during various types of military activities? This paper seeks to explore these and other critical issues such as:

(a) The ethical roles and responsibilities of taking part in military deployments
(b) Whether doctors who take part in offensive, combat operations could return to the practice of medicine and still be regarded as ethical medical practitioners when they are not involved in combat operations
(c) Whether special rules apply to health care practitioners in conflict and combat situations
(d) And if indeed so, what these rules should be and how they should be applied?

The author believes that it is the duty and role of ethics experts, scholars and communities such as Military Medical Ethics Committee to provide decisive guidance on ethical practice in the military communities. Countries that have medics deployed externally have experienced situations that present moral ambiguity. If all those cases were brought to light how many of those medics would still be allowed to practice today. If indeed we say that certain military activities for medics are barred, who would then perform these functions?

1.3 *Methodology*

Mixed methods of literature review and interviews of former Special Forces operators in the same unit as Dr. Basson were used for this paper. The literature review looked into impartiality, neutrality and focused on grey areas of medical neutrality. Interviews were with former colleagues that worked closely with Dr. Basson or shortly after him in the same unit. Most interviewees gave the information on anonymity which is understandable because the case of Dr. Basson is extremely divisive. These were all military personnel who worked with Dr. Basson during various phases. Some worked with him as doctors in a healthcare setting treating patients, while others worked with Dr. Basson in a Special Forces and research environment where no patients were involved.

2 Results

This section discusses the findings from literature review, interviews with colleagues of Dr. Basson. It also discusses and makes comments about Dr. Basson's criminal trial, HPCSA hearing and ethical issues that arose from thereon.

2.1 *Special Forces Operations Pertinent to Dr. Basson*

In the 1980s South African Medical Service under the South African Defence Force (SADF) was slightly different than today: the medical battalion where Dr. Basson was a project director was detached to Special Forces. Only in 1994, the South African Military Health Service (SAMHS) became an independent combat service support that provides medical support, limited nuclear, chemical and biological defence support to the army, air force and the navy (Defence Web 2023).

The South African Special Forces' work remains highly classified hence the secrecy of their operations. Due to the strategic nature of special operations, the brigade is an autonomous force, and its operations are controlled at the highest level to ensure security. Key capabilities and specialized skills enable the effective conduct of operations. This may happen under severe conditions, and this is the area that military practitioners like Dr. Basson functioned and operated.

Regular military operations are vastly different from Special Forces operations. Special Forces operators are required to be independent for 72 hours, with no evacuation or assistance available, often required to stabilize and care for injured colleagues or patients while fending off fire or attack. These situations are challenging and fraught with ethical challenges and dilemmas. Health care operators are armed and are part of the combat team.

Doctors can only deploy on special military operations if they have the same military training as the group that is being supported, especially in Special Forces' operations where a relatively small group is moving behind enemy lines (Blunden 2023). Fully trained by Special Forces, the medics will take part in all group activities including carrying some of the group's combat equipment, in addition to most of the medical equipment and food supplies. Furthermore, if the group hits a contact, the doctors will then take part fully in the resulting firefight in defence of the group. The doctor here may be required to act like a full combatant soldier. If they are successful, this would mean that the doctor may have killed some of the attacking enemy troops. These doctors know that they cannot claim protection as a medical officer under the Geneva Convention, and have accepted that if captured, they will be treated as ordinary Prisoners of War. Can these doctors return to practicing medicine after such deployments?

Within a Special Forces group, health personnel may not wear any identifiers depending on the nature of the operation (Blunden 2023). This would mean no country uniform, dog tags or identity documents that would confirm them as doctors. In these situations, the practitioner is a combatant and soldier. The ethical question here is whether this individual is still bound by full medical ethics? Can the practitioner be accepted as a doctor or a soldier with medical knowledge?Moreover, Special Forces doctors will have very limited medical resources and may sometimes have to choose to treat wounded own force members before more severely wounded enemy soldiers. Though in principle and in law doctors remain neutral and should provide medical care for all including enemy patients, it is said that the ordinary doctor's ethics stays in you (Blunden 2023).

2.2 *The HPCSA Hearing of Dr. Basson*

Despite the not guilty verdict at the criminal trial of Dr. Basson, the HPCSA in 2007, launched an investigation into perceived/alleged ethical issues/transgressions flowing from the criminal trial. The HPCSA, in its investigation, considered matters which Basson's lawyers did not dispute in court. In 2013 after lengthy legal issues the HPCSA found Dr. Basson guilty of unprofessional conduct on four of the charges, as follows:

Charge 1: Dr. Basson coordinated the production of large quantities of illegal psychoactive substances.

Charge 2: Dr. Basson was involved in equipping mortars with teargas for use against Angolan government soldiers.

Charge 3: Dr. Basson provided SADF operatives with disorienting substances to facilitate illegal kidnapping.

Charge 4: Dr. Basson made available cyanide capsules to SADF soldiers so that if captured they could decide to commit suicide to avoid revealing military information under pressure.

The HPCSA disciplinary committee convicted Dr. Basson using the evidence that was uncontested during his criminal trial. The committee concluded that those acts constitute a breach of medical ethics and are not trivial and should be dealt with accordingly (HPCSA 2014). This conclusion shed light on an ethical dilemma and heightened controversy around the doctor with some feeling that the basic principle of first *do no harm* was clearly violated and felt that the sanction against Dr. Basson should match the conviction, some even called it egregious (Vula 2023). Dr. Basson claimed that the disciplinary action by HPCSA was as a result of the small group of doctors who had a vendetta against him (Vula 2023), this is largely due to the few healthcare professionals that lodged the initial complaints against Dr. Basson in 2000 and those other HCPs that subsequently signed on to that submission (Vula 2023).

2.3 Dr. Basson's Arguments and Refutations at the HPCSA Hearing

Dr. Basson's main defence at the HPCSA hearing was that there was no doctor-patient relationship during these activities, and that he acted in his capacity as a soldier and not as a doctor. This can be supported by the statement from the interview that at the time of Project Coast Dr. Basson was a combatant and an operator subservient to Special Forces Command and therefore had to act and take orders like all other Special Forces operators. Dr. Basson told the committee: "As a medical professional and an ethical person I was happy with the fact that I was sure that I was saving lives and in that way that I would not damage any individual" (Parker 2012). Dr. Basson's nine main arguments and refutations raised at the hearing are as follows (as summarised in Vula 2023):

(a) *The circumstances of war justified extreme measures.* The council rejected this argument and reminded Dr. Basson that medical ethics are identical in war and in peace times.
(b) *The respondent was under military orders.* The council also rejected this argument saying that a medical doctor is responsible as an individual for his or her actions.
(c) *Dr Basson was acting as a soldier and not a doctor.* The council rejected this argument and said that Dr. Basson ought to have deregistered as a doctor and should never have switched roles between being a doctor and being a soldier. The council also added that his role could not have been done by someone who was not a doctor; he was specifically recruited because he was a medical professional. He was recruited for his knowledge, skills and privileges of a doctor.
(d) *Military ethics are different than peacetime ethics and military doctors do not have autonomy.* Basson's witnesses quoted an Israeli ethicist, a non-doctor, who argued that normal principles for medical reasoning are absent in military situations, doctors do not have autonomy. The committee rejected this argu-

ment, pointing to numerous international guidelines that make it clear that there cannot be two standards.

(e) *There was no doctor-patient relationship, so no ethical restraints needed.* The council rejected this argument and said that some international guidelines do not restrict themselves to patients, but speak of victims for example, the Declaration of Tokyo (WMA 2016). The council also mentioned that where doctor's actions impact on people, they must take responsibility for their decisions and behaviour, even when there is no explicit doctor-patient relationship.

(f) *Dr Basson was a young inexperienced doctor so cannot be held responsible.* The council rejected this argument. In Dr. Basson's case, he was not inexperienced when taking on the role as head of the CBW programme; in fact, he was not a young follower but a mature leader. Therefore, he ought to take responsibilities for his actions.

(g) *Dr Basson was not aware of international codes and guidelines.* The council rejected this argument saying that "ignorance cannot be bliss". No professional recruited to a job within a CBW programme can avoid checking ethical and legal legitimacy when accepting the position.

(h) *Ethical standards changed and were not the same in the 1980s.* While the council acknowledged that ethical emphasis changes over time, it said that there was enough ethical guidance in the 1980s to have conveyed very clear concern about this practice.

(i) *Chemical substances developed were designed to be non-fatal.* The council found that Dr. Basson used his medical skills for the purpose of abetting military operations, and that this was not compatible with professional standards, nor was supplying military operatives with suicide tablets. The council's disciplinary committee came to the decision that these violations conflicted with medical ethics and represented unprofessional conduct. The court could not determine that he provided soldiers with suicide pills and Dr. Basson never admitted to producing suicide pills like cyanide. He argued that the substances he produced were non-fatal.

A challenge arises from the fact that two organs of state, the SANDF and HPCSA put the health care practitioner between a rock and a hard place: Dr. Basson is a good example of this conflict stemming from dual loyalty, the grass on which two bulls are fighting (Chatugah 2007). Some of the things that Dr. Basson was accused of are still being done today as extrapolated from the interviews conducted. Military medical personnel are there to fulfil their duty and there is nothing more to it. Dr. Basson oversaw a particular project to be used by Special Forces. Some of these systems are still being utilised to this day within the prescripts of CBW Convention. Hence the question, should special rules apply to military medical personnel? Should these individuals be allowed to practice after leaving Special Forces operations? There are no clear answers.

3 Brief Overview of Medical Ethics Aspects of the Case

Medical ethics deal with the rights and wrongs of decision making in clinical practice (Markose et al. 2016). Medical Ethics are principles-based Ethics, stemming from Principlism popularised by Tom Beauchamp and James Childress (2001). Principlism encompasses: Beneficence, Non-Maleficence, Autonomy and Justice Principles.

Non-Maleficence: This principle asserts an obligation not to inflict harm intentionally and forms the framework for the standard of due care to be met by any HCP. However, there are plenty of examples of wrongdoing by HCPs from literature e.g. Tuskegee Syphilis which is a trial that was carried out by the US Public Health Service from 1932 (Tyson 2000). The trial was only stopped in 1972 when an advisory panel found it unethical (Tyson 2000). This is one principle that is repeatedly mentioned as being breached by Dr. Basson.

Autonomy: Autonomy brings together many ethical challenges, including the areas of consent and capacity. Having unwilling participants in research or clinical trial is seen as coercion. The other charge in Dr. Basson's case was that he ran experiments on soldiers, this is assumed that it was without their informed consent. And the treatment he had hoped to administer to the victims of Project Coast would have been without their consent. This charge was later withdrawn though by the HPCSA.

Justice: This is the principle of fairness. It refers to the equitable, and appropriate treatment in light of what is due or owed to person. Dr. Basson's actions may have contravened the principle of justice as seen by HPCSA and those calling for sanctions as a way of administering justice on behalf of those he may have harmed.

3.1 HPCSA Hearing and Ethical Issues in Dr. Basson's Case

One concern is the potential conflict between the duty of medical professionals to prioritize the health of their patients and their obligation to serve the strategic interests of their military commanders. Numerous articles have been published regarding military medical personnel compelled to breach patient confidentiality or provide interventions/ treatments that are not medically necessary in order to gather intelligence, interrogate detainees, or support military operations (Childress 2013). Such actions violate the fundamental ethical principle of *primum non nocere* (first do no harm). They undermine trust between physicians and patients, threaten the integrity of the profession, and breach the ethical foundations of the profession. The case of Dr. Basson was alleged as an example of such, though no doctor-patient relationship existed.

Another concern is the issue of medical care as a tool for coercion or punishment. In some cases, military medical personnel may be asked to provide medical care to detainees as a means of inducing them to collaborate with interrogators or military officials. This practice violates the fundamental ethical principle of non-maleficence first do no harm principle, demands that medical care should be provided solely for the benefit of the patient above all considerations. This is further discussed in the section pertaining to the charges against Dr. Basson by the HPCSA hearing.

Moreover, there are broader questions about the relationship between health and national security that arise when medical care is used to further non-medical ends. South Africa has a number of examples of such cases during apartheid. Some of the military operations require medics and medical equipment for various reasons, such as medical care to the deployed forces, treatment of the wounded, defence against or use of chemical, biological, radiation and nuclear attack (Raeeszadeh et al. 2022). To achieve strategic objectives and win wars or have the upper hand in conflicts; the use of medical resources as a tool of war can be seen as obscuring the primary mission of medicine, which is to promote health, alleviate suffering and not deliberately cause harm or kill.

When medical personnel are perceived as serving the interests of the military or state rather than promoting the health of the people, it can undermine the trust and credibility of the medical profession. Even though Dr. Basson was exonerated by the courts and the guilty finding by the profession's council was set aside by the Supreme Court of Appeal of South Africa, the perception across the world makes his case a perfect example of such cases in history.

This section looks at applicable ethical principles in Dr. Basson's case as argued by the HPCSA. The HPCSA hearing was convened because the standard burden of proof in criminal matters in South Africa is proof beyond reasonable doubt (Helman and Peuker 2021). However, this is not the level required in determining professional unethical conduct. According to this statement it therefore means that even though Dr. Basson was acquitted in his criminal trial, his professional medical ethics were clearly flawed (HPCSA 2014).

Charge one, which states that Dr. Basson coordinated the production of large quantities of illegal psychoactive substances, is in contravention to medical ethics. For example, the World Medical Association's (WMA) Tokyo Declaration (first adopted in 1975, and revised lastly in 2016) states that a doctor *"shall not provide any premises, instruments, substances or knowledge to facilitate the practice of torture or other forms of cruel, inhuman or degrading treatment or to diminish the ability of the victim to resist such treatment"*. Hence the HPCSA found Dr. Basson guilty of contravening the principle of non-maleficence.

The second charge states that Dr. Basson was involved in equipping mortars with teargas for use against Angolan government soldiers. Teargas is classified as a chemical weapon by the Geneva Convention and the use teargas against enemy soldiers in combat is prohibited. *"Better to shoot them than teargas them"* commented one interviewee. Despite that, private laboratories develop chemical weapons according to CB convention system and produce CB agents in small samples for testing PPE suits' effectiveness in protecting against biological weapons (Montazzolli 2021). This also is in contradiction with the principle of non-maleficence and physicians shall not participate in forms of inhuman procedures, whatever the offence of which the victim of such procedure is accused of, according to the WMA (Declaration of Tokyo, 2016).

The third charge which states that Dr. Basson provided SADF operatives with disorienting substances to facilitate illegal kidnapping. This is also contravening the principle of non-maleficence. Though it is argued that is it better to shoot and kill the person? As opposed to kidnapping if it is to ensure the safety of one's country from illegal, perhaps deadly activities. Snatching has been used before, for example by Israel as previously mentioned in the introduction, using an unnamed sedative that rendered him semi-conscious, in order to make the snatching successful (Estrin 2019). This is what is also referred to in the case of Dr. Basson in relation to Project Coast.

The fourth charge which states that Dr. Basson made available cyanide capsules to SADF soldiers so that they could decide to commit suicide to avoid revealing military information or the whereabouts of fellow soldiers under torture. This could be considered as having contravened the autonomy of the soldiers, their individual capacity for intentional action or Freedom to Choose. Dr. Basson also argued that doctors have no autonomy. HPCSA dropped some of the charges, including one related to using troops and police task force members as guinea pigs for sedatives such as mandrax and ecstasy as well as incapacitating drugs and teargas in the 1980s (Masombuka 2012). These charges relate to members' autonomy and according to the accusations it would mean that the soldiers and police were given no choice to consent.

The handling of the case by HPCSA raises its own ethical questions, for example the wording in their article on their website posted 09 July 2014 suggests some level of biasness and not impartiality for example the opening sentence read *The HPCSA is elated at the verdict of the Professional Conduct Committee that found Dr Wouter Basson guilty of unprofessional conduct during his time as Head of Project Coast* (HPCSA 2014). The more balanced reporting as an independent professional body should have reported on the outcome of the inquiry without being seen to be picking sides. However, this kind of wording is what is expected from such a body as the HPCSA, for example the chairperson of Dr. Basson's inquiry into his unethical conduct quoted from the WMA that *medical ethics in time of armed conflict is identical to medical ethics in time of peace.*" Furthermore *"the primary obligation of the physician is his professional duty; the physician's supreme guide is his conscience.* (HPCSA 2014). This was an appropriate thing to say from a body such as the HPCSA.

3.2 Main Ethical Arguments against Dr. Basson

Supporting arguments for striking Dr. Basson off the roll go around the fact that Dr. Basson never applied for amnesty with the Truth and Reconciliation Commission, has never apologized for what he did and has shown no respect for the moral choices expected of a doctor (Njanji 2014). They argue that many doctors during apartheid faced difficult moral choices which included a consideration of several issues including human rights, and this happened at high cost to themselves (Njanji 2014). Some doctors these choices cost them their lives (Njanji 2014). They also suggest that Dr. Basson's actions subjugated the ethics of medicine to political interests. An argument made by O'Malley (1999) that the interests of the patients were thus frequently subjugated to those of the State.

Those arguing for Dr. Basson's harsher sentence gave an example of the doctors that treated Steve Biko who allowed security forces to cloud their clinical judgement and helped to cover up his torture and subsequent death (Jenkins and McLean 2004). Steve Biko was a leader of the Black Consciousness Movement in South Africa, died in 1977 while being detained by security police (Jenkins and McLean 2004). One of these doctors, Dr. Tucker was subsequently struck off the register as a result, hence they are calling for Dr. Basson to also be struck off the register though their cases are not quite similar. One case had no known patients involved and in the other case there is a known patient who died under their care.

This group against Dr. Basson also claims that the public is being misled through misinformation. They are claiming that news information on this case is unbalanced and mainly comes from Dr. Basson's friends and some of his patients. They say that the public is still unaware of the full extent of Dr. Basson's activities and perhaps it is in the public interest for that information to be made known. Perhaps the next step is to openly engage with Dr. Basson to offer full explanation for immunity. It would be foolish to expect full truth because of the allegiance to Special Forces but less sensitive information can be shared. In the author's view it would be sufficient to provide closure. Taking him through another council's enquiry may diminish chances of the whole truth ever coming out. A lot of time has passed therefore, Dr. Basson may be persuaded to talk openly if promised guaranteed immunity.

Dr. Basson's detractors say that chemical and biological warfare is inhumane and should never be fuelled by doctors, hence the adoption of the two international conventions and humanitarian laws to eliminate these weapons. This is in line with the four Geneva Conventions of 1949 (UN 2003). They also mention that these weapons are still being allowed to be used in countries like Syria (Nuclear Threat Initiative 2018). This means that there are doctors and scientists who continue to support military forces to develop such weapons. Their intention is about being deliberate about taking a stand against this heinous military action which is in violation of humanitarian laws (Reynolds 2015).

They are also arguing that, even though it took 13 years to reach a decision, HPCSA followed due process. The medical council gave Dr. Basson ample opportunity to clear his name, the fact that Dr. Basson did not conduct himself in a manner comfortable with his detractors, is taken as a confirmation that his actions were very unethical (Reynolds 2015). They claim that he has cultivated an image of a caring doctor but has simply refused to take responsibility for his unethical actions. They argue that therefore, he should not continue to practice and he has pretended that the disciplinary action is the result of a vendetta against him by a minority of doctors plus other interested parties (Baleta 2012).

3.3 *The Current Situation of Dr. Basson's Hearing at the HPCSA*

The acting president of the Supreme Court of Appeal, found in favour of Basson. His ruling said Basson had contended that a penalty imposed by a committee remains effective until the appeal is finalised. This situation, Basson argued, is prejudicial to him, in that such penalty would cause irreparable harm to his practice (Timeslive 2018). This was 5 years after he was found guilty of unethical conduct by the HPCSA. The reason given by the court to overturn the medical council's findings was because of bias on the part of some of the presiding committee members and ruled that the proceedings were irregular, unfair and disregarded Dr. Basson's rights; and that the finding of unethical conduct be set aside (Timeslive 2018).

As a result of this and various other legal proceedings the 2013 recommendation by the HPCSA that Dr. Basson be removed from the medical register is moot. Dr. Basson is still in private practice pending any further actions by the HPCSA. In 2021, the HPCSA confirmed that it had drawn up a new charge sheet and planned to start proceedings against Dr. Basson again in the near future (Nkosi 2021). To date, this has not happened yet. This may be perceived as unjust by Dr. Basson and those who support him as this HPCSA matter has now dragged on for over 15 years.

4 Discussion

Should Dr. Basson be struck off the medical register? Can Dr. Basson's patients sue the HPCSA at a later stage if he is struck off the roll; if the medical council allowed him to treat patients for over 30 years after Project Coast? The reason for suing the medical council would be that if they knew all along that Dr. Basson was not fit to treat patients but allowed him to continue treating patients 30 years after the Project Coast had ended. The question really is what would striking him off the register achieve at this point? What about the well-being of some of his patients where he is their only cardiologist? What is the alternative punitive action to being struck off the register? Is it even fair now to subject him to another medical council hearing? What would that mean for other Special Forces doctors throughout the world? Those interviewed telephonically and in person spoke about how he cared for all his patients equally. They say as a doctor to his patients he is remarkable and so far, there is no evidence to point to the contrary.

The ethical issue of whether military doctors can dissociate their medical and military personas has important implications. The HPCSA's provisional findings seem to indicate that the answer is no. Neutrality of medical officer in Special Operations requires deep investigation on what is happening, according to one of the people interviewed for this paper. This should not be an issue for military hospital doctors or those supporting conventional military forces. But what about Special

Forces doctors who, when deployed, carry combat equipment and are expected to take an active part in any firefight in defence of the group? This remains unanswered.

Other views seem to suggest that, Special Forces doctors should be allowed to practice as medical doctors after Special Forces military operations. This is because they possess the necessary qualifications, skills and experience to provide valuable medical services to society. Special Forces doctors undergo rigorous training and are often exposed to high-pressure and demanding environments during military operations. They are trained to operate in challenging conditions, with limited resources, and provide medical care to soldiers and civilians in combat zones or disaster-stricken areas. This unique experience equips them with specialized medical knowledge, quick decision-making abilities, and the ability to work efficiently under intense circumstances.

After their service in Special Forces, these doctors can bring a range of invaluable skills and perspective into their civilian medical practice. Lots of resources were spent on these doctors, it would be a waste to society to stop them from practicing. Their experience in trauma surgery, emergency medicine, and field medicine can greatly benefit patients facing critical injuries or requiring urgent medical attention. Moreover, their familiarity with improvisation, resource management, and the ability to work in chaotic situations can enable them to effectively handle challenging situations that may arise in civilian medical settings. This makes them invaluable for the public sector especially. Furthermore, Special Forces doctors often work in diverse cultural and social contexts, gaining exposure to unique healthcare practices, diseases, and approaches to treatment. This exposure can enhance their overall medical knowledge, allowing them to provide more comprehensive care to patients. It is important that Special Forces doctors undergo any necessary reintegration or transitional training if required for civilian practice. This would ensure that they are up-to-date with the latest advancements in medical science, regulations, and ethical practices unique to civilian healthcare. There ought to be a deliberate intent from medical councils in order to achieve this.

In summary, allowing Special Forces doctors to continue their medical careers after military operations enable them to utilize their specialized skills and experiences for the benefit of society. Their proficiency in delivering healthcare under extreme conditions and their diverse perspectives can greatly enhance the quality of medical care available to the general public.

If the issue of non-dissociation of Special Forces doctors' medical and military personas is upheld these doctors cannot return to practicing medicine ever again after such deployments. This has important implications, even if there are no clear answers on whether special rules must be applied to these doctors.

5 Conclusion

Application of ethics in military medical practise is complex and controversial and this leaves the practitioners exposed to possible sanctions by their statutory bodies. The protracted nature of Dr. Basson's prosecution at both the criminal courts and

HPCSA highlights the difficulty of finding neutrality and harmonising dual loyalties for military healthcare practitioners. This case remains prominent in modern history where medical means were used for non-medical ends.

If issues of non-dissociation of their medical and military personas is upheld by the HPCSA these doctors cannot return to practising medicine ever again after such deployments. This leaves the questions surrounding Special Forces medical operators after service unanswered. This paper explored these questions but did not answer all of the issues asked, as this is a complex case. Moreover, due to the length of time it has taken to address and finalise the case of Dr. Basson at the medical council, may be infringing on Dr. Basson's right to a fair inquiry and justice.

Providing medical care to further non-medical ends is a practice that raises significant ethical and moral concerns in military medical ethics. While there may be circumstances in which the use of medical resources to support military operations is justifiable, care must be taken to ensure that the ethical principles of medicine are not compromised or undermined. Military medical professionals have an obligation to prioritize the health of their patients and to ensure that medical care is provided in a manner that is consistent with the fundamental values of the profession. By upholding these principles, military medicine can serve both the healthcare needs of individuals and the strategic intents of the state.

A possible recommendation learned from the case is that alternative group of independent ethicists involving military medical doctors should be consulted to advise HPCSA on the way forward just to bring back the credibility into the case. Included in this group should be doctors with Special Forces experience, both serving and retired, and non-medical ethicists in order to present a balanced view. This will also bring much needed closure to the case. It will perhaps answer the questions that this case review failed to provide.

5.1 Note

> What is not disputed is that Dr Basson is a competent cardiologist who continues to run a successful practice, there have been no complaints by patients against him that cast doubt upon his competency or that since entering private practice he has not adhered to ethical principles (Lotter and Nienaber 2020).

References

Adeaga, Favour. 2022. *Wouter Basson bio: Age, family, education, book, profiles, net worth*. https://briefly.co.za/94146-wouter-basson-bio-age-family-education-book-profiles-net-worth.html. Accessed 12 August 2023.

Baleta, Adele. 2012. Apartheid doctor to face south African disciplinary hearing. *World Report* 379: 604. https://doi.org/10.1016/S0140-6736(12)60254-2.

Beauchamp, Tom L., and James F. Childress. 2001. *Principles of biomedical ethics*. 5th ed. New York: Oxford University Press.
Biological Weapons Convention. 1972. *United Nations' Office for disarmament affairs*. https://disarmament.unoda.org/biological-weapons/. Accessed 8 Aug 2023.
Blakemore, Erin. 2019. *The Daring Israeli Spy operation to capture Nazi Mass Murderer Adolf Eichmann*. https://www.history.com/news/adolf-eichmann-nazi-capture-holocaust-trial-mossad. Accessed 5 Aug 2023.
Blunden, Chris. 2023. Col (Dr) interviewed for this paper, a former colleague of Dr Wouter Basson.
Chatugah, Tigere. 2007. *When two bulls fight, it is the grass that suffers*. https://envirocom.wordpress.com/2007/01/24/when-two-bulls-fight-it-is-the-grass-that-suffers/#:~:text=The%20age%2Dold%20African%20saying,of%20Zimbabwe's%20ailing%20economy%20%E2%80%93%20literally. Accessed 8 Aug 2023.
Childress, Sarah. 2013. *Report: Medical professionals participated in torture of terror suspects*. https://www.pbs.org/wgbh/frontline/article/report-medical-professionals-participated-in-torture-of-terror-suspects/. Accessed 8 Aug 2023.
Defence Web. 2023. *A guide to SANDF*. https://www.defenceweb.co.za/wp-content/uploads/Repository/A-Guide-to-the-SANDF/A-guide-to-the-SANDF-chapter-12-samhs.pdf. Accessed 8 Aug 2023.
Estrin, Daniel. 2019. *The Doctor who helped Israeli Spies catch Eichmann but refused recognition for it*. https://www.npr.org/2019/07/16/740686212/the-doctor-who-helped-israeli-spies-catch-eichmann-but-refused-recognition-for-i. Accessed 5 Aug 2023.
Health Professions Council of South Africa Corporate Affairs 2014. Dr Wouter Basson Found Guilty of Unprofessional Conduct https://www.hpcsa-blogs.co.za/dr-wouter-basson-found-guilty-of-unprofessional-conduct/. Accessed 20 Mar 2023.
Helman, Lawrence and Andrea Pekeur. 2021. *Legal systems in South Africa: Overview*. https://uk.practicallaw.thomsonreuters.com/w-030-7871?transitionType=Default&contextData=(sc.Default)&firstPage=true. Accessed 5 Aug 2023.
Hlati, Okuhle. 2021. Pressure piles on Mediclinic over apartheid-era doctor Dr Wouter Basson. Pressure piles on Mediclinic over apartheid-era doctor Dr Wouter Basson. https://www.iol.co.za/capetimes/news/pressure-piles-on-mediclinic-over-apartheid-era-doctor-dr-wouter-basson-7477407f-e1a9-4ff7-bea6-8b112f5644db. Accessed 28 July 2023.
Jackson, Miles. 2015. A conspiracy to commit genocide: Anti-fertility research in Apartheid's chemical and biological weapons programme. *Journal of International Criminal Justice* 13: 933–950. https://doi.org/10.1093/jicj/mqv060.
Jenkins, T., and G.R. McLean. 2004. The Steve Biko affair. *Medicine, Crime, and Punishment* 364. https://doi.org/10.1016/S0140-6736(04)17634-4
Lotter, R., and Annelize Gertruida Nienaber. 2020. *Twenty years an accused. An analysis of two decades of proceedings against doctor Wouter Basson*. Available at: http://hdl.handle.net/2263/82760. Accessed 16 May 2023.
Marbot, Olivier. 2021. *South Africa: 'Dr Death' discovered to still be practising medicine*. https://www.theafricareport.com/63661/south-africa-dr-death-discovered-to-still-be-practising-medicine/ Accessed 12 Aug 2023.
Markose, Aji, Krishnan Ramesh, and Ramesh Maya. 2016. Medical ethics. *Journal of Pharmacy and Bioallied Sciences* 8: 1–4. https://doi.org/10.4103/0975-7406.191934. Accessed 30 June 2022.
Masombuka, Sipho. 2012. *Basson fights charges*. https://www.timeslive.co.za/politics/2012-01-24-basson-fights-charges/. Accessed 6 Aug 2023.
Montazzoli, Matt. 2021. *Surrounding battlefield use of tear gas*. https://www.justsecurity.org/76595/clearing-the-fog-of-war-surrounding-battlefield-use-of-tear-gas/. Accessed on 5 Aug 2023.
Njanji, Tinashe. 2014. *Accountability for apartheid's agents: Why Dr Wouter Basson deserves to be struck off the medical register*. https://www.phm-sa.org/accountability-for-apartheids-agents-why-dr-wouter-basson-deserves-to-be-struck-off-the-medical-register/. Accessed 20 Mar 2023.

Nkosi, Bongani. 2021. *HPCSA has still not served Dr Death his new charge sheet*. https://www.iol.co.za/the-star/news/hpcsa-has-still-not-served-dr-death-his-new-charge-sheet-24080753-a9b2-4044-a5de-a032fba8ec61. Accessed on 5 Aug 2023.

Nuclear Threat Initiative. 2018. *Syria country spotlight*. https://www.nti.org/countries/syria/. Accessed 8 Aug 2023.

O'Malley, Padraig. 1999. *The heart of hope*. https://omalley.nelsonmandela.org/index.php/site/q/03lv01508/04lv01509/05lv01510.htm. Accessed on 5 Aug 2023.

Parker, Faranaaz. 2012. *Basson: I did it for the benefit of mankind*. https://mg.co.za/article/2012-03-26-basson-i-did-it-for-the-benefit-of-mankind/. Accessed 6 Aug 2023.

Raeeszadeh, Mohammad, Goodarzi Hassan, Moghaddam Jamal Akhavan, Machian Mohamad Sadegh Bahaghighat, Gooshki Hossein Shamsi, and Khoshi Aabolfazl. 2022. The role of military medicine and the significance of training military medicine. *Journal of Family Medicine and Primary Care* 8: 4184–4189. https://doi.org/10.4103/jfmpc.jfmpc_80_22. Epub 2022 Aug 30.

Reynolds, Louis. 2015. *Why Wouter Basson should be removed from the medical register*. https://www.dailymaverick.co.za/opinionista/2015-01-21-why-wouter-basson-should-be-removed-from-the-medical-register/. Accessed 20 Mar 2023.

South African Government. 1997. *The Constitution of the Republic of South Africa*. https://www.gov.za/documents/constitution/constitution-republic-south-africa-1996-1. Accessed on 21 Mar 2023.

Swart, Mia. 2008. The Wouter Basson prosecution: The closest South Africa came to Nuremberg? *Heidelberg Journal of International Law* 68: 209–226.

TimesLive. 2018. *Basson wins challenge on professors deciding his sentence for misconduct*. https://www.timeslive.co.za/news/south-africa/2018-01-17-basson-wins-challenge-on-professors-deciding-his-sentence-for-misconduct/. Accessed 20 Mar 2023.

Tyson, Peter. 2000. Holocaust on trial. The Experiments. https://www.pbs.org/wgbh/nova/holocaust/experiside.html%20. Accessed on 9 June 2022.

United Nations. 2003. *International norms and standards relating to disability*. https://www.un.org/esa/socdev/enable/comp210.htm#:~:text=The%20four%20Geneva%20Conventions%20of,no%20active%20part%20in%20hostilities. Accessed 28 July 2023.

Vula Online Content. 2023. *A statement of concern from Health professionals in South Africa: Dr Wouter Basson*. https://vula.uct.ac.za/access/content/user/01378260/WB%20Petition.htm. Accessed 25 Mar 2023.

World Medical Association. Declaration of Tokyo—Guidelines for physicians concerning torture and other cruel, inhuman or degrading treatment or punishment in relation to detention and imprisionment, originally adopted in 1975 and lastly revised in 2016. https://www.wma.net/policies-post/wma-declaration-of-tokyo-guidelines-for-physicians-concerning-torture-and-other-cruel-inhuman-or-degrading-treatment-or-punishment-in-relation-to-detention-and-imprisonment/#:~:text=The%20physician%20shall%20not%20countenance,and%20in%20all%20situations%2C%20including

Chapter 11
Medical Neutrality in Times of Military Coup in Myanmar

Wunna Tun

Abstract On February 1, 2021, Myanmar had a military coup. Arresting and kiilling health care professionals, attacking hospitals and clinics, the barbaric slaughter of thousands of innocent people, particularly women and children, are all routine in Myanmar today. There have been several incidents of atrocities of Myanmar military and security forces against our medical professionals. Since the Myanmar military and security is purposely bombing civilians and hospitals in Myanmar, violating the neutrality of health care in active battle, Myanmar's health staff have been mobilized under exceptionally difficult conditions. Over three years, tens of thousands of our medical education colleagues and students have been prisoned and some being killed. The medical professionals were arrested and killed because of providing medical treatment to injured. Medical neutrality is often disrespected in Myanmar: There are several statements by international medical community for Myanmar. The Myanmar military and security forces disregard them and more and more Myanmar health professionals suffer and even family members of medical professionals have been arrested and killed when they cannot find medical personnel. International NGO are reluctant to provide humanitarian assistance to refugees and injured because their staff have greatest chance of endanger their own life as well as their family members due to Myanmar military action. This chapter discusses medical neutrality from the perspective of a humanitarian health care provider in Myanmar during military coup and critically reflect on own experience on medical neutrality in actual combat zones in Myanmar.

It is generally accepted that medical facilities must not be used as military objectives, and medical workers must not be targeted in both peace and war time (WMA 2017). Medical personnel should never be persecuted, penalized, murdered, or wounded for carrying out their ethical responsibilities to give treatment. It should adhere to Geneva Conventions in order to ensure that physicians and other health

W. Tun (✉)
International Medical Education Academy for Myanmar Health Professionals, Undisclosed Location, Myanmar

D. Messelken, A. E. Barbar (eds.), *Challenging Medical Neutrality*, Military and Humanitarian Health Ethics, https://doi.org/10.1007/978-3-031-69398-4_11

care professionals may offer care to everyone in need during times of armed war. All efforts must be taken to guarantee that patients and victims have complete access to healthcare (WMA 2017). Physicians must have unrestricted access to patients, medical facilities, and equipment, as well as the necessary protection to carry out their professional duties. Doctors and healthcare workers must not be stopped from carrying out their professional responsibilities. Interfering with or penalizing doctors who provide medical treatment to patients violates international law.

Medical neutrality has been referred to in various WMA, ICRC and International humanitarian law, namely their responsibilities under the 1949 Geneva Conventions, notably its common article 3, the obligations applicable to them under the 1977 Additional Protocols thereto 2005, and the legal standing of them is not applicable in these. One of the goals of the WHO Global Health for Peace Initiative, which was announced at the 75th World Health Assembly in 2022, is to "influence conflict dynamics in a positive way," which violates the concept of medical neutrality while adding to the strain of an already overburdened staff.

The remainder of this article will focus on medical neutrality in the real world: What's Happening in Myanmar. I would like to mention that these sources are reporting events in this article that I have witnessed myself (in a similar way) and that I have also heard about from trusted colleagues.

Myanmar is the country in South east Asian bounded by global power of neighboring countries. According to Myanmar Health Statistics 2020, there are 54 million population resided in Myanmar of which two third of them are in rural. However, the total doctor to population ration is 7-53 doctors per 100000 in which most doctor is denser in urban area.

Every day, we personally witness how the military coup ruins our lives—our homes, our relationships, our very existence. Medical textbook theories, images and stories from novels, movies, and television shows about medical ethics, war, and military coups have ceased to be fiction and have instead become reality. 1.6 million people in Myanmar are becoming totally internally displaced persons due to Myanmar military and security forces (UNHCR 2023; UNOHCA 2023). There are nearly 30000 civilians' properties have been burnt due to Myanmar military. There are 18.6 million Myanmar people who need humanitarian assistant including 10 million people who desperately need of health care while more than 15 million Myanmar are hunger from food security and 5.4 million are shortage of education.

The same tales have rippled through many health professionals in Myanmar. On the first day of the coup, we rushed to our hospital to help the injured. Myanmar military blocked injured people from obtaining treatment, and hospitals themselves have become killing fields. We set up a clinic to treat patients and train physicians and students in the city, however arrests are subject to discretion. We tried to teach using an internet portal which is also being pursued by the Myanmar military. We simply bring ourselves out of our comfort zone and engage in activities we would not typically do. Finally, they were able to retreat to the different part of Myanmar borders, where we created field clinics that treated everyone who came in. We train health professionals to be surgical assistants since the nearest hospital with an operating room was a day's ride away by motorbike along a muddy road, out of reach for the majority of people.

Millions of people in Myanmar are without power (only diesel generators and solar energy), safe running water, an internet connection, and a home to call their own (France-Presse 2021) (WeAre[WP] and Assistweb 2022), (Nachemson 2021) Infection outbreaks and a spike in injured patients are worsened by food and water shortages, displacement, and infrastructure and health-care system disruption. As a result of the continued attacks on medical professionals and civilians, psychological consequences are on the rise.

Myanmar experienced a military coup on February 1, 2021 (Krishna and Howard 2021). From February 1, 2021 until the Russian invasion of Ukraine, Myanmar accounts for more than 75% of worldwide healthcare attacks. Following the Russian invasion, we have had an equivalent number of attacks on healthcare statistics with our Ukrainian colleagues. Arresting medical personnel, bombing hospitals and clinics, and slaughtering thousands of innocent people, mainly women and children, are all commonplace in Myanmar today. There are several incidences of Myanmar military abuses against our medical staff. Due to deliberate bombings of citizens and hospitals by the Myanmar military and security forces, which violates the neutrality of healthcare in times of war, the country's medical personnel have been forced to mobilize under extremely trying circumstances. Tens of thousands of our coworkers and students in the health professions have been imprisoned and some have even been killed during the past three years.

Arrest warrants and other legal actions have been taken against physicians and nurses for their medical care. In addition to the indiscriminate shelling of civilian populations, there is the extensive burning and destruction of civilian dwellings and infrastructure, including clinics and hospitals. According to the Johns Hopkins Bloomberg School of Public Health report, there are numerous evidence of Myanmar Military and security forces attacking hospitals and clinics with indiscriminate artillery and arson, attacking medical staff, continuing raids and occupations of these facilities, arresting large numbers of health workers and re-arresting recently released health workers, multiple instances of prisoners dying in detention as a result of medical neglect, and increasing attacks on local non-governmental organizations (NGOs), particularly those that aid displaced populations in ethnic minority areas.

In Myanmar, medical neutrality is frequently disregarded. Because medical professionals treated injured people with medicine, they were slain and jailed. Myanmar's military and security forces disregarded statements, resolutions made against Myanmar by the international medical community. When they cannot obtain help, more and more medical professionals in Myanmar suffer, and even their relatives have been detained and killed. Since their personnel commitment has the best chance of survival, international NGOs are hesitant to give humanitarian aid to refugees and the injured.

1 Arbitrary Arrest of Health Care Service Providers or Individuals Seeking Medical Care

Throughout the 3-year military coup, the junta has continued to arrest and detain doctors and nurses who provide medical care in private clinics and hospitals (Seventy-eighth session Agenda item 71, 2023). Since the military coup in 2021, arbitrary arrests of health care providers have been a common practice of Myanmar's military and security forces. The security forces have detained medical professionals at their homes and workplaces, suspended the credentials of well-known doctors, searched hospitals for injured resistance fighters, and threatened to shut down medical centers (Paddocks 2022). Once they learn that the person is a doctor, they are free to detain and question the doctor as they like. The following are some examples of military brutality against detained health care providers. The military stormed the biggest hospital in Myanmar and arrested civilians Professor who are in the midst of providing medical care to injured patients in accordance with Physician Pledge (ACEM 2021). While arrested, he was in hospital managing injured patients which are caused by Myanmar military and security forces (Joint statement – Calling for urgent release of professor Maw Maw Oo 2021a, b). He refuted the military order not to treat these kinds of patients while maintaining medical neutrality. The medical professionals intended to provide assistance to individuals in need, but in the middle of the route, the junta soldiers found out, followed them, and started shooting. A nurse who was treating the internally displaced people was shot. (Root and Lusan 2023).

The Myanmar military also arrested doctors and nurses who had been offering medical attention, first aid instruction, medical supplies (The Irrawaddy 2023d). The people were also detained by the authorities for organizing the delivery of medical supplies to IDPs (The Irrawaddy 2023a). They had confiscated more than 160 packages of consumable medical products and medical supplies (The Irrawaddy 2023b). The medical professionals escaped to a remote region to Myanmar while their colleagues were being imprisoned treating patients with general ailments as well as those injured by military bombardment and land mines in improvised tent clinics in camps (Milko 2022).

Myanmar health workers have been killed in attacks on people, such as air strikes and shelling, or while providing medical care to injured civilians. During the crackdown, health workers were detained or injured, ambulances and medical equipment were confiscated, and buildings were destroyed. The fate of most of the detained health workers remains unknown (Safeguarding Health in Conflict Coalition 2023). Every day, new names are added to the regime's "indictment list of doctors," and arrests of those charged continue (Dyer 2021).

2 Arbitrary Arrest of Family Member of Health Care Service Provider or Individual Seeking Medical Care When They Cannot Find Medical Personnel

The Myanmar military junta has arbitrarily arrested and detained family members of healthcare workers and individuals seeking medical care as captives, subjecting them to torture, insufficient food and water, and other inhumane conditions in a systematic violation of medical neutrality principles aimed at punishing and intimidating medical professionals and civilians. Since the coup, Myanmar soldiers have routinely utilized people as human shields, continuing a decades-long pattern of abuse by Myanmar's military (Seventy-eighth session Agenda item 71 2023). The Myanmar health professionals treated the Myanmar insurgents and paid with their lives (Service 2022). Myanmar military and security forces often arrested together their family members as well as family members of medical professionals when they did not find health professionals as a barbaric act of ransom so that they can handcuff themselves to them. The junta in Myanmar kidnaps the family members of medical professionals who are wanted as hostages and locks the child, parent and friend in captivity. The detained medical professionals and their family members were not given enough food in detention and had to bathe in toilet water (Coconuts Yangon 2021). They even burned down the homes of family members of health workers and displaced family members (Root and Lusan 2023).

3 Myanmar Military Attack on Individuals in the Course of Receiving Medical Treatment

Myanmar military and security forces made it difficult for health workers, humanitarian groups, and local NGOs to visit patients and populations in conflict-affected areas, further restricting humanitarian space (Safeguarding Health in Conflict Coalition 2023). The Myanmar military disregards fundamental principles of medical neutrality and attacks individuals seeking medical care. During a raid on a church in eastern Kayah State, troops found 48 patients receiving care there, seven of whom had Covid-19 as well, then made the arrests of both the patients and 18 medics (Reuters 2021) One of the most dangerous places in the world for a health worker to work today is in Myanmar, as patients are treated in secret locations. Patients are reluctant to seek medical care for fear of endangering themselves (Barber 2022).

The Myanmar military opened fire with heavy artillery and ammunition, killing medical personnel who were on the front lines trying to treat wounded compatriots (Firn 2023). People seeking treatment in covert clinics are referred to by the regime as "illegal patients" as it pursues injured civilians (Paddock 2022). Many people have fled to informal bush settlements with little access to medical care and food as Myanmar military and security forces has robbed displaced populations of medical and humanitarian assistance. Pharmacies must obtain clearance from Myanmar military and security forces before selling drugs to avoid losing their licenses (Safeguarding Health in Conflict Coalition 2023). Myanmar health-care personnel have been shot or abused while treating those in need, with troops sometimes lying in wait to shoot on those who help the injured (Dyer 2021). Myanmar's coup dictate punishes doctors with shootings, arrests, and cancellation of their licenses to practice, and denounces the Myanmar passport (Dyer 2021).

4 Myanmar Military Attacks on Health Care Facilities

Myanmar military and security forces have carried out shootings and armed raids on hospitals. Myanmar's military is invading educational and medical institutions to establish a base and set up camp (Beech 2021a, b). Using a map of public hospitals and clinics that it has compiled in recent years, the Myanmar military junta has deliberately targeted medical facilities and medical workers with airstrikes, military incursions, and the burning of structures (Maryland Coordination and Analysis Center 2023). These attacks were carried out with the express purpose of terrorizing the population. These hospitals, which provide health care to the neighborhood, were deliberately targeted by the junta. There was no active armed fighting there. It was a deliberate attack and bombed the structure that houses a hospital and a rest home (Asiannews 2022). Some of the hospitals have been targeted by junta bombings several times in a month. The first airstrike damaged part of the hospital. The entire structure was destroyed in another attack (Irrawady 2023b). A refugee camp housing some IDPs was attacked by junta helicopter gunships and jets, killing many medical staff, civilians and children (Irrawady 2023c). Another evidence that the junta bombed the hospital facility, both citizens and refugees received medical care. The hospital has provided medical care to at least 300 war refugees (Pike 2023). A nearby refugee camp was also attacked. The hospital was devastated by the junta's bombing, which mainly provided maternity care to various communities. From the military attack on the hospital facility, the X-ray machine was destroyed (Nora 2023).

5 Malicious Obstruction of Access to Medical Facilities and Personnel

Myanmar health workers have been physically assaulted and arrested while providing care. The Myanmar military and security forces violate medical neutrality by obstructing access to medical facilities and personnel through military occupation, closing hospitals, and revoking licenses to practice (Firm and Reed 2023). The military violates medical neutrality by halting the provision of free medical care and demanding that private hospitals and clinics turn over patients' identities and medical histories in an effort to coerce physicians into working in facilities it controls (Paddocks 2022). Medical professionals who are apprehended by the military or police face increased danger of torture and abuse in detention as authorities seek information about their underground networks that provide medical treatment to people in need in accordance with the physician's pledge. The junta has significantly reduced the quality of medical care in areas near Myanmar's borders, resulting in fragmented clinics with limited access to supplies and facilities (Tatum 2022). Health facilities are frequently used by Myanmar military and security forces to gain military advantage or to imprison people (Johns Hopkins Bloomberg School of Public Health 2021). The cruel actions of Myanmar's military and security forces have led to an increase in attacks on medical personnel providing care to civilians. A medic's body was eventually discovered with signs of a close-range execution (Johns Hopkins Bloomberg School of Public Health 2021).

6 Malicious Obstruction of Access to Medicine

Myanmar's military and security forces restrict aid distribution to vulnerable communities, despite the country's significant humanitarian needs (Seventy-eighth session Agenda item 71 2023). Military personnel have barred access to aid and medical supplies. After the military takeover in 2021, the junta and its security forces tightened travel restrictions on aid workers, obstructed assistance convoys and access highways, and destroyed non-military goods and medicines (Regan 2023). The ongoing obstruction of roads and telecommunications in conflict areas of Myanmar is impeding the ability of health professionals to act in compliance with the principle of medical neutrality. Since the military took over, the generals have cruelly refused to provide life-saving aid to anyone affected by the fighting, ostensibly as a form of retribution (Wionweb Team 2021). The oxygen that many people needed to breathe has been snatched by the military leaders in Myanmar. The security personnel seized oxygen supply and barred people from going to hospitals for immediate care. They detained humanitarian workers who tried to bring oxygen to the afflicted areas (Leng 2021). The military has ordered the withholding of oxygen from private clinics and has even banned charities from distributing it. The military has also prohibited citizens from purchasing supplies from oxygen producers, causing desperate family members to rescue ailing relatives. The troops even shot into a gathering of civilians waiting in line to purchase oxygen tanks (Beech 2021a, b). At a large number of security checkpoints along important land routes in Myanmar, Junta soldiers seize medicine (Ground news 2023). No one can move the medicine under these difficult circumstances. People have to hide the medicine they bring, and it takes about a month to deliver the medicine. Transportation is stopped even if it contains paracetamol or something similar (Milko 2022). The children and the elderly in Myanmar become malnourished and develop subsequent health problems as a result of the malicious obstruction of food and medical supplies.

The military regime in Myanmar is preventing foreign aid agencies from providing food and medicine to those most in need (Strangio 2023). They have imposed further restrictions on mobility and banned international organizations from entering certain areas. These organizations have many beneficiaries in these areas (Rashid 2023). Because of the junta's confiscations at road checkpoints, humanitarian organizations are no longer able to deliver food and medicine to displaced citizens (Ground News 2023). The regime's soldiers have been killing people while disobeying medical neutrality. At a hospital that was treating locals and displaced people, the military and security forces of Myanmar also set fire to equipment and medications valued at millions of dollars. Medical supplies were stolen, and hospital equipment was discarded in wells. A ventilator and an ambulance were destroyed during the attack (The Irrawaddy 2023d). Drivers delivering medicines have been detained, abused, and their supplies taken (Safeguarding Health in Conflict Coalition 2023). In Myanmar, a variety of local groups are responsible for delivering humanitarian aid, including medical supplies. These organizations face increasing obstacles to providing needed supplies and services, including damage to offices and raids on supplies, as well as arrests and violence against staff. Attacks on local nongovernmental organizations (NGOs), particularly those providing assistance to displaced persons in ethnic minority areas, have increased (Johns Hopkins Bloomberg School of Public Health 2021).

7 Conclusion

Healthcare workers in Myanmar face significant stress due to insufficient resources and violent situations. Doctors, medical students, and nurses and their family members in Myanmar are being arrested, killed, revoked their license to practice medicine, harassment, or intimidation for providing critical care to people in need during the military coup. Maintaining medical neutrality allows health workers to treat patients freely, independently, and ethically without fear of punishment or intimidation. Myanmar's military forces use health facilities, personnel, and patients as human shields, forcing health workers to choose between abandoning patients and risking their own lives during fighting. It is essential to respect the insignia and ensure that health workers and facilities are not misused for military purposes or targeted attacks. As the Myanmar military has blocked the supply chain, the international community should provide essential medicines and equipment directly to the affected community through the border areas to ensure that all patients have unhindered and safe access to health care. We call on the international community to ensure that all violations of international treaties protecting health workers in Myanmar are promptly and impartially investigated and that the perpetrators of these war crimes are brought to justice, including before the International Criminal Court.

International laws, resolutions, and statements that fail to translate into actual action and sanctions for violations of medical neutrality fuel the desire of Myanmar's military and security forces to interrogate, assassinate, or otherwise target health care workers and facilities. Health professionals in Myanmar value life and are committed to upholding medical ethics in accordance with the Physician's Pledge (Revised WMA Declaration of Geneva or A modern successor to the Hippocratic Oath) and International Codes of Medical Ethics, and to maintaining medical neutrality regardless of their origin or affiliation. During Myanmar's darkest period, Myanmar medical professionals have dedicated their lives to protecting and healing the injured and caring for those in need. We offer hope that all things are possible if we continue to do our best in every way.

References

AsiaNews.it. 2022. *Church and convent bombed in demoso*. MYANMAR. Available at: https://www.asianews.it/news-en/Church-and-convent-bombed-in-Demoso-55324.html. Accessed 30 Sept 2023.

Barber, H. 2022. Secret health service: Myanmar's military junta drives medics underground. *The Telegraph*. Available at: https://www.telegraph.co.uk/global-health/terror-and-security/secret-health-service-myanmars-military-junta-drives-medics/. Accessed 30 Sept 2023.

Beech, H. 2021a. As Covid rages in Myanmar, army hoards oxygen, doctors say. *The New York Times*. Available at: https://www.nytimes.com/2021/07/15/world/asia/myanmar-covid-oxygen.html. Accessed: 1 Oct 2023.

———. 2021b. *Myanmar military storms universities and hospitals and revokes press licenses. The New York Times*. Available at: https://www.nytimes.com/2021/03/08/world/asia/myanmar-coup-violence.html. Accessed 4 Oct 2023.

Coconuts Yangon. (ed.). 2021. *7-year-old son of Myanmar health official still behind bars as calls for release intensify*. Coconuts. Available at: https://coconuts.co/yangon/news/7-year-old-son-of-myanmar-health-official-still-behind-bars-as-calls-for-release-intensify/. Accessed 30 Sept 2023.

Dyer, O. 2021. Myanmar coup regime targets striking doctors with bullets, arrests, and erasures. *BMJ: British Medical Journal* 373: n1076. https://doi.org/10.1136/bmj.n1076.

Firn, M. (ed.) 2023. Junta Shell Kills Medic in Myanmar's Sagaing Region. *Asia Barta*. Available at: https://asiabarta.com/english/2023/02/06/junta-shell-kills-medic-in-myanmars-sagaing-region/. (Accessed: 30 September 2023).

Firn, M., and M. Reed (eds.). 2023. *Myanmar's junta shuts down 3 Mandalay Hospitals*. Radio Free Asia. Available at: https://www.rfa.org/english/news/myanmar/mandalay-hospitals-closed-05182023051613.html. Accessed 1 Oct 2023.

France-Presse, A. 2021. *Coup-hit Myanmar sees nationwide electricity outage*. NDTV.com. Available at: https://www.ndtv.com/world-news/coup-hit-myanmar-sees-nationwide-electricity-outage-2384292. Accessed: 27 Sept 2023.

Ground News. 2023. Myanmar junta's confiscation of food, medicine leaves Kayah State residents at risk. *Ground News: Breaking News Headlines and Media Bias*. Available at: https://ground.news/article/myanmar-juntas-confiscation-of-food-medicine-leaves-kayah-state-residents-at-risk. Accessed 1 Oct 2023.

Johns Hopkins Bloomberg School of Public Health. 2021. *Violence against health care in Myanmar*. [online] Physicians for Human Rights. Available at: https://phr.org/our-work/resources/violence-against-health-care-in-myanmar/.

Joint statement – Calling for urgent release of professor Maw Maw Oo. 2021a. *ACEM*. Available at: https://acem.org.au/News/April-2021/JOINT-STATEMENT-%e2%80%93-calling-for-the-urgent-release-o. Accessed 27 Sept 2023.

———. 2021b. *ACEM*. Available at: https://acem.org.au/News/April-2021/JOINT-STATEMENT-%e2%80%93-calling-for-the-urgent-release-o. Accessed 27 Sept 2023.

Krishna, G. and S. Howard. 2021. Myanmar doctors are under fire from the military and covid-19. *BMJ: British Medical Journal* [Preprint]. https://doi.org/10.1136/bmj.n2409.

Leng, S.H. 2021. *Myanmar Military arrested aid workers. SHAN News (Burmese)*. Available at: https://burmese.shannews.org/archives/25443. Accessed 4 Oct 2023.

Maryland Coordination and Analysis Center. (ed.). 2023. *MCAC, Myanmar military junta 'deliberately bombing medical facilities' with secret jungle hospital now being sought*. Available at: https://mcac.maryland.gov/2023/09/myanmar-military-junta-deliberately-bombing-medical-facilities-with-secret-jungle-hospital-now-being-sought/. Accessed 1 Oct 2023.

Milko, V. 2022. Death, danger, despair: A year in Myanmar under the military. *AP News*. Available at: https://apnews.com/article/myanmar-one-year-into-military-rule-7cd0ee904a358cb230525ff0d8a0becd. Accessed 1 Oct 2023.

Nachemson, A. 2021. Why is Myanmar's military blocking the internet? *Censorship News | Al Jazeera*. Available at: https://www.aljazeera.com/news/2021/3/4/myanmar-internet-blackouts. Accessed 27 Sept 2023.

Nora. 2023. Myanmar junta increasingly targeting healthcare: Nug. *The Irrawaddy*. Available at: https://www.irrawaddy.com/news/burma/myanmar-junta-increasingly-targeting-healthcare-nug.html. Accessed 30 Sept 2023.

Paddock, R.C. 2022. *Myanmar's health system is in collapse, 'obliterated' by the regime. The New York Times*. Available at: https://www.nytimes.com/2022/04/19/world/asia/myanmars-coup-doctors.html. Accessed 1 Oct 2023.

Pike, J. 2023. *Junta helicopter drops bomb on hospital where war refugees were being treated*. Available at: https://www.globalsecurity.org/wmd/library/news/myanmar/2023/myanmar-230425-rfa01.htm. Accessed 30 Sept 2023.

Rashid, M. 2023. Medical charity faces access restrictions in Myanmar. *The Irrawaddy*. Available at: https://www.irrawaddy.com/news/burma/medical-charity-faces-access-restrictions-in-myanmar.html. Accessed 1 Oct 2023.

Regan, H. 2023. *Myanmar junta travel restrictions are holding up vital aid to cyclone-hit communities. CNN*. Available at: https://www.cnn.com/2023/05/18/asia/cyclone-mocha-aid-restrictions-myanmar-intl-hnk/index.html. Accessed 4 Oct 2023.

Reuters. 2021. Myanmar troops arrest 18 medics for treating members of anti-junta groupsR. *CNN*. Available at: https://www.cnn.com/2021/11/24/asia/myanmar-arrest-medics-intl-hnk/index.html. Accessed 30 Sept 2023.

Root, R.L. and N.N. Lusan. 2023. In Myanmar's worsening conflict, health workers deliver care and Dodge death. *The New Humanitarian*. Available at: https://www.thenewhumanitarian.org/news-feature/2023/04/12/myanmar-conflict-health-workers#:~:text=From%20February%202021%20to%20September%202022%2C%20750%20health,%28SAC%29%20forces%20or%20to%20the%20junta-run%20police%20force. Accessed 30 Sept 2023.

Safeguarding Health in Conflict Coalition. 2023. *Insecurity Insight Data on People in Danger*. [online] Available at: https://insecurityinsight.org/wp-content/uploads/2023/05/SHCC-Report-Ignoring-Red-Lines.pdf.

Service, J.H. and B. 2022. *She treated Myanmar's resistance and paid with her life*. BBC News. Available at: https://www.bbc.com/news/world-asia-62047569. Accessed 30 Sept 2023.

Seventy-eighth session Agenda item 71 (c) Promotion and protection of human rights: human rights situations and reports of special rapporteurs and representatives Situation of human rights in Myanmar Note by the Secretary-General*. 2023. Available at: https://daccess-ods.un.org/access.nsf/Get?OpenAgent&DS=A/78/527&Lang=E. Accessed 18 June 2024.

Strangio, S. 2023. Myanmar military restricts access to areas hit by Cyclone Mocha. *The Diplomat*. Available at: https://thediplomat.com/2023/06/myanmar-military-restricts-access-to-areas-hit-by-cyclone-mocha/. Accessed 4 Oct 2023.

Tatum, M. 2022. Escalating threats to health workers in Myanmar. *The Lancet* 399 (10325): 619. https://doi.org/10.1016/s0140-6736(22)00270-7.

The Irrawaddy. 2023a. Myanmar regime seizes 5 medics for 'supplying resistance fighters'. *The Irrawaddy*. Available at: https://www.irrawaddy.com/news/burma/myanmar-regime-seizes-5-medics-for-supplying-resistance-fighters.html. Accessed 30 Sept 2023.

———. 2023b. Regime airstrike destroys hospital in Lower Myanmar. *The Irrawaddy*. Available at: https://www.irrawaddy.com/news/burma/regime-airstrike-destroys-hospital-in-lower-myanmar.html. Accessed 30 Sept 2023.

———. 2023c. The worst Myanmar regime airstrikes on civilians. *The Irrawaddy*. Available at: https://www.irrawaddy.com/news/burma/the-worst-myanmar-regime-airstrikes-on-civilians.html. Accessed 30 Sept 2023).

———. (ed.). 2023d. Myanmar Junta destroys medicine and Food Stores for the displaced in Sagaing. *The Irrawaddy*. Available at: https://www.irrawaddy.com/news/burma/myanmar-junta-destroys-medicine-and-food-stores-for-the-displaced-in-sagaing.html. Accessed 1 Oct 2023.

UNHCR. 2023. *Myanmar UNHCR displacement overview 18 Sep 2023, UNHCR Operational Data Portal (ODP)*. Available at: https://data.unhcr.org/en/documents/details/103606. Accessed 19 Sept 2023.

UNOCHA. 2023. *Myanmar, Front page*. Available at: https://www.unocha.org/myanmar. Accessed 19 Sept 2023.

WeAre[WP] and Assistweb. 2022. *Myanmar: Water shortage amidst economic, political and Health Crisis*. SOLIDARITÉS INTERNATIONAL. Available at: https://www.solidarites.org/en/live-from-the-field/myanmar-water-shortage-amidst-economic-political-and-health-crisis/. Accessed 27 Sept 2023.

WION Web Team. 2021. Myanmar junta blocking life-saving aid from reaching millions: Report. WION. Available at: https://www.wionews.com/south-asia/myanmar-junta-blocking-life-saving-aid-from-reaching-millions-report-436704. Accessed 4 Oct 2023.

WMA. (ed.). 2017. WMA – The World Medical Association-Ethical Principles of Health Care in times of armed conflict and other emergencies. The World Medical Association. Available at: https://www.wma.net/policies-post/ethical-principles-of-health-care-in-times-of-armed-conflict-and-other-emergencies/. Accessed 20 Sept 2023.

Chapter 12
Economic Sanctions Policy, Medical Neutrality and the Human Right to Health

Julian W. März

Abstract Economic sanctions are a key tool of foreign policy, which former UN Secretary-General Kofi Annan has described as a "necessary middle ground between war and words". Economic sanctions are, however, often criticized for their humanitarian impact and their potential collateral damage to health and healthcare, which can potentially go far beyond the target country: If sanctions disrupt global supply chains of alimentary, pharmaceutical and medical products, they can potentially entail collateral damage not only in the target country, but also in third countries, particularly those with strained and less resilient healthcare systems. In this chapter, I will analyse the implications of the human right to health and the principle of medical neutrality for the use of economic sanctions and will present possible solutions to mitigate the unintended humanitarian impact of economic sanctions.

The principle of medical neutrality imposes ethical obligational on all parties involved in armed conflicts. Chapters 3–8 and 11 have analysed the ethical obligations imposed on physicians active in armed conflict settings. Obligations imposed on States and military institutions by virtue of the principle of medical neutrality have been addressed in Chapters 5–11. The present chapter will discuss the implications of the principle of medical neutrality within a broader human rights and policy context focusing on the topic of economic sanctions.

Economic sanctions are a key tool of foreign policy which former United Nations Secretary General Kofi Annan has described as "a necessary middle ground between war and words" (UN 2022a). Economic sanctions are a key instrument (to attempt) to enforce respect for international law, including the Charter of the United Nations, the Geneva Conventions and (more broadly) international human rights and humanitarian law. At the same time, economic sanctions are often fiercely criticised for the

J. W. März (✉)
Institute of Biomedical Ethics and History of Medicine (IBME), University of Zurich, Zurich, Switzerland
e-mail: julian.maerz@ibme.uzh.ch

D. Messelken, A. E. Barbar (eds.), *Challenging Medical Neutrality*, Military and Humanitarian Health Ethics, https://doi.org/10.1007/978-3-031-69398-4_12

unintended health impacts they can have and are sometimes portrayed as human rights violations themselves. The present chapter will, in particular, illustrate potential violations of the human right to health and the principle of medical neutrality by economic sanctions, and provide policy recommendations how to design and implement economic sanctions in compliance with human rights and medical neutrality requirements.

1 Economic Sanctions Policy of the United Nations

At the global level, the United Nations Security Council has established a total of 31 sanctions regimes since 1966 (as of August 2023) (UNSC 2023). Some sanctions regimes have targeted states and state government officials, e.g., those against Iran, North Korea and Mali (which ended in August 2023). Other sanctions regimes have targeted armed groups (e.g., the Houthi rebels in Yemen, military coup leaders in Guinea-Bissau and criminal syndicates in Haiti) and international terrorist organizations (e.g., ISIL, Al-Qaida and the Taliban). In addition, the United Nations Security Council has imposed arms embargoes on several politically unstable countries and countries in civil wars, e.g., Lebanon, Libya and the Democratic Republic of the Congo (UN Sanctions App 2023).

These sanctions regimes vary widely in scope, purpose and content. One of the widest-reaching sanctions regimes has been established with regard to the Democratic People's Republic of Korea in order to enforce respect of the Treaty on the Non-Proliferation of Nuclear Weapons. As the nuclear weapon program is the key priority of North Korean politics, the Security Council has imposed a broad range of sanctions on North Korean officials and key industries in an attempt to cut funding and to induce policy change. For instance, the Security Council has passed resolutions banning the trade of military equipment, freezing assets of political leaders and government officials, restricting fishing rights, restricting scientific and technical cooperation, capping labour exports, and banning the export of certain goods (e.g., fish and seafood) used by the North Korean government to procure foreign currencies (CFR 2022).

Sanctions regimes can fulfil a variety of purposes. For sanction regimes in force as of September 2023, the UN sanctions database identifies the purposes listed in Table 12.1.

These sanctions regimes imposed by the United Nations Security Council are complemented by sanctions at the regional and national level. In some cases, e.g., those of Belarus, Myanmar, Russia, and Syria, veto rights in the United Nations Security Council have blocked the adoption of international sanctions regimes, despite evidence of severe breaches of international human rights and humanitarian law. In the case of Mali, a shift in global politics has led to the discontinuation of a UN Security Council sanctions regime established in 2017, which ended in August 2023. In these cases, regional and national sanctions, e.g., by the United States and the European Union, have been imposed outside of UN Security Council sanctions regimes.

Table 12.1 Aims of UN Security Council Sanctions Regimes in Force as of Sep 30, 2023 (based on UN Sanctions App 2023)

Country/ Organization	In force since...	Measure (Examples)	Purposes
Al-Qaida, ISIL and associates	1999	Travel ban, asset freeze, arms embargo, aviation ban, ban on chemical imports for heroin processing and Iraqi and Syrian cultural property exports	Counter-terrorism
Central African Republic	2013	Travel bans and asset freezes on 12 individuals and 2 entities; arms embargo	Peacebuilding, human rights protection
Democratic People's Republic of Korea	2006	Travel bans, asset freezes, arms embargo, trade embargo targeting many economic sectors (e.g., mining, aviation, finance)	Non-proliferation of nuclear weapons
Democratic Republic of Congo	2003	Travel bans and asset freezes on 36 individuals and 9 companies, arms embargo	Peacebuilding, human rights protection
Guinea-Bissau	2012	Travel bans on 11 individuals	Democracy support
Haiti (UN 2022b)	2022	Travel ban, asset freeze and arms embargo on individuals designated by a special UN committee (1 listed as of Sep 2023)	Reestablishing peace, safety and respect for human rights
Iran	2006	Travel ban, asset freeze, arms embargo, embargo on sensitive technology and investment	Non-proliferation of nuclear weapons
Iraq	2003	Asset freeze and arms embargo on 86 individuals and 61 companies	Non-proliferation of nuclear weapons, peacebuilding
Lebanon	2005	Arms embargo	Counter-terrorism, peacebuilding
Libya	2011	Arms embargo, aviation ban, travel ban on 28 individuals, asset freeze on 23 individuals and 2 companies	Peacebuilding, democracy support, human rights protection
Somalia	1992	Travel ban and asset freeze on 15 individuals and 1 entity, arms embargo	Counter-terrorism, peacebuilding, human rights protection
South Sudan	2015	Arms embargo, travel ban and asset freeze	Peacebuilding, human rights protection
Sudan	2004	Arms embargo, travel ban and asset freeze	Peacebuilding, democracy support, human rights protection
Taliban	2011	Arms embargo, travel ban and asset freeze	Counter-terrorism, human rights protection
Yemen	2014	Arms embargo, travel ban and asset freeze	Peacebuilding, democracy support, human rights protection, support of humanitarian efforts

In 2002 and 2014, the United Nations General Assembly (UNGA) and the United Nations Human Rights Council (UNHRC) have passed non-binding resolutions condemning economic sanctions not covered by UN Security Council resolutions (UNGA 2002; UNHRC 2014). Despite this, national and regional sanctions regimes remain much-used instruments of foreign policy.

2 Economic Sanctions and Human Rights

Economic sanctions have often been criticised for the potentially devastating impact on the civilian population of the sanctioned country they can have. Table 12.2 summarizes the potential negative consequences of economic sanction regimes on the population of the target country.

The compatibility with international human rights law and medical ethics is a key condition for the acceptability of economic sanctions, as underlined by the United Nations Committee on Economic, Social and Cultural Rights:

Table 12.2 Potential Effects of Economic Sanctions on the Population of the Sanctioned Country (based on UNICEF 2022 and Pintor et al. 2023)

	Direct effects	Indirect effects
Economic	Increase in inflation rates, increase in prices of food and medicines, higher unemployment, lower household income	Increase in child labour, school dropout, government spending cuts to education, healthcare and social security, deprioritization of environmental protection, funding cuts to HIV prevention campaigns (Aldebron, 1993)
Health, water and sanitation	Unavailability of essential medicines and medical equipment, decline in water, sanitation and hygiene infrastructure, shortage in raw materials for essential goods and services	Decrease in life expectancy, increase in malnutrition, worse learning outcomes, increasing spread of infectious diseases (see Gibbons and Garfield 1999; Barry 2000; and Aloosh and Aloosh 2023)
Food	Unavailability of food imports and materials for agriculture (e.g., fertilizers), reduced government funding of food supply	Increase in child labour, lower breastfeeding rates
Education	Unavailability of school equipment (e.g., education software), barriers to international cooperation of researchers, barriers of access to scientific/medical literature (Butler 2019)	Long-term economic impact, increase in women's and children's HIV rates (see Kim 2019a, b)
Pandemic response (e.g., COVID-19)	Unavailability of medicines, medical supplies and protective equipment (see Murphy et al. 2020), limits to online education and telemedicine	Increased socio-economic impact of a pandemic, higher risks to vulnerable groups

In adopting this general comment the sole aim of the Committee is to *draw attention to the fact that the inhabitants of a given country do not forfeit their basic economic, social and cultural rights by virtue of any determination that their leaders have violated norms relating to international peace and security.* The aim is not to give support or encouragement to such leaders, nor is it to undermine the legitimate interests of the international community in enforcing respect for the provisions of the Charter of the United Nations and the general principles of international law. *Rather, it is to insist that lawlessness of one kind should not be met by lawlessness of another kind which pays no heed to the fundamental rights that underlie and give legitimacy to any such collective action.* (CESCR 1997)

Particularly in the case of wide-reaching sanctions, potential conflicts with various human rights, e.g., the rights to food, to water and sanitation, and to health, and principles of medical ethics, e.g., solidarity, justice, and medical neutrality, have to be considered when designing the sanctions regime. In the following, I will focus particularly on the implications of the human right to health and the principle of medical neutrality for economic sanctions policy and implementation.

The principle of medical neutrality imposes an obligation of non-interference with medical services during armed conflicts or other humanitarian crises (Carruth 2020). The World Medical Association (WMA) *Declaration on the Protection and Integrity of Medical Personnel in Armed Conflicts and Other Situations of Violence* states that all parties involved in situations of violence have an obligation to "enable healthcare personnel to treat injured and sick patients (…), to ensure that safe access to adequate medical facilities for the injured and others in need of medical aid is not unduly impeded (…), and to ensure that the equipment including personal protection equipment necessary for the safety of healthcare workers is available to them as needed" (WMA 2022).

The human right to health is enshrined in various international law sources. For instance, the preamble of the Constitution of the World Health Organisation (1946) states:

> Health is a state of complete physical, mental and social well-being and not merely the absence of disease or infirmity. *The enjoyment of the highest attainable standard of health is one of the fundamental rights of every human being without distinction of race, religion, political belief, economic or social condition.* The health of all peoples is fundamental to the attainment of peace and security and is dependent upon the *fullest co-operation of individuals and States.* The extension to all peoples of the benefits of medical, psychological and related knowledge is essential to the fullest attainment of health. (WHO 1946)

Article 25 of the Universal Declaration of Human Rights also proclaims the human right to health.

> *Everyone has the right to a standard of living adequate for the health and well-being of himself and of his family, including food, clothing, housing and medical care and necessary social services, and the right to security in the event of unemployment, sickness, disability, widowhood, old age or other lack of livelihood in circumstances beyond his control.* Motherhood and childhood are entitled to special care and assistance. All children, whether born in or out of wedlock, shall enjoy the same social protection. (UNGA 1948)

A guarantee of the human right to health is enshrined in Article 12 of the International Covenant on Economic, Social and Cultural Rights (ICESCR) (1966) and Article 24 of the Convention on the Rights of the Child (CRC) (1989).

Art. 12 ICESCR:

> 1. *The States Parties to the present Covenant recognize the right of everyone to the enjoyment of the highest attainable standard of physical and mental health.*
> 2. The steps to be taken by the States Parties to the present Covenant to achieve the full realization of this right shall include those necessary for: (a) The provision for the reduction of the stillbirth rate and of infant mortality and for the healthy development of the child; (b) The improvement of all aspects of environmental and industrial hygiene; (c) The prevention, treatment and control of epidemic, endemic, occupational and other diseases; (d) The creation of conditions which would assure to all medical service and medical attention in the event of sickness.

Art. 24 CRC:

> 1. States Parties recognize the right of the child to the *enjoyment of the highest attainable standard of health and to facilities for the treatment of illness and rehabilitation of health.* States Parties shall strive to ensure that no child is deprived of his or her right of access to such health care services.
>
> (…)
>
> 4. *States Parties undertake to promote and encourage international co-operation with a view to achieving progressively the full realization of the right recognized in the present article. In this regard, particular account shall be taken of the needs of developing countries.*

The United Nations Committee on Economic, Social and Cultural Rights (CESCR) considers health a "*fundamental human right indispensable for the exercise of other human rights*" (CESCR 2000). It considers that the right to health imposes on states an obligation to "*refrain at all times from imposing embargoes or similar measures restricting the supply of another State with adequate medicines and medical equipment*" and underlines that "*restrictions on such goods should never be used as an instrument of political and economic pressure*" (CESCR 2000).

3 Health Impact of Economic Sanctions

Economic sanctions can impact health in various ways. The present chapter proposes to differentiate between three types of health impact: direct, indirect and collateral (Table 12.3).

In some cases, economic sanctions are targeting (at least incidentally) health services or healthcare providers. Examples can include export restrictions on dual use products in the medical, pharmaceutical and life sciences sector or the freezing of assets of (e.g., military) healthcare providers. In these cases, the impact of the sanctions on healthcare is felt directly by the population of the sanctioned country. In some cases, the health of healthcare professionals can also be seriously put at risk if, for instance, economic sanctions are impeding the import of protective equipment (e.g., FFP-2 masks against COVID-19).

Table 12.3 Examples of the three types of health impact of economic sanctions

Direct	Unavailability of medicines in the sanctioned country due to a trade embargo on the medical, pharmaceutical and life sciences sector or on (dual use) raw materials Unavailability of medicines due to the impossibility to mobilize foreign currency assets held in bank accounts subject to asset freezes Unavailability of healthcare due to a freezing of assets of (e.g., military) healthcare providers Restrictions to "medical tourism" by citizens of the sanctioned country
Indirect	Economic consequences of sanctions leading to reduced state income entailing spending cuts to the healthcare sector Disengagement of pharmaceutical, life sciences and finance companies from the sanctioned country due to fears of violations of sanctions or reputational damage Emigration of healthcare professionals due to loss of income and loss of opportunities in the sanctioned country
Collateral	Disruption of global medicines supply chains Disruption to global trade, economic impact on third countries Cuts to development aid and humanitarian assistance to third countries by the sanctioned country due to loss of state income

The main health impacts of economic sanctions are, however, regularly only felt indirectly by the population of the country targeted by the sanctions. Sanctions can, for instance, increase operational risk and reputation risk and dissuade pharmaceutical companies to do business in or with a sanctioned country. If pharmaceutical companies are ceasing business operations in the sanctioned country, this can potentially lead to severe shortages in medicinal products. The same effect can, however, also occur if banks or logistics companies are ending business operations in or with a sanctioned country. More broadly, the economic impact of economic sanctions can lead to spending cuts to healthcare in the sanctioned country, which can also entail serious negative population health impacts. Furthermore, economic sanctions can also provoke brain drain and a loss of training opportunities and access to medical research findings in the sanctioned country which can compromise the quality of healthcare services (Bezuidenhout et al. 2019). Deteriorations of population mental health can be linked, e.g., to rising unemployment rates and economic hardship caused by economic sanctions (Aloosh et al. 2019). Road traffic injuries and deaths also seem to be significantly more common in countries during periods of sanctions, which might be due to lower levels of investment in road safety and trauma care (Duric and Peek-Asa 2008).

Finally, economic sanctions can also have a health impact which goes far beyond the sanctioning and the sanctioned country. The potential impact on the global economy and global trade can lead to economic effects, but also direct effects on healthcare, in third countries, particularly low- and middle-income countries. This can include disruptions to global pharmaceutical, food and energy supply chains, but also economic consequences which can have direct impacts on health care provision. Therefore, the African Union has expressed concern over potential disruptions to grain supplies to Africa due to EU and US sanctions on Russia, particularly on the banking and finance sector (Financial Times 2022).

The effects of economic sanctions are often particularly felt by vulnerable populations (e.g., children, persons with disabilities or chronic diseases), who lack the resources and power to compensate income losses and to overcome additional barriers of access to healthcare. Similarly, persons living with rare diseases and neuropsychiatric and oncological diseases are particularly vulnerable to the incidental effects of economic sanctions (see Shahabi et al. 2015; Deilamizade and Esmizade 2015; Aloosh 2018; and Asadi-Pooya et al. 2022). Addressing these effects is a shared responsibility of the sanctioning and the sanctioned state, as emphasized by the UN Committee on Economic, Social and Cultural Rights:

"In considering sanctions, it is *essential to distinguish between the basic objective of applying political and economic pressure upon the governing élite of the country to persuade them to conform to international law, and the collateral infliction of suffering upon the most vulnerable groups within the targeted country*. (…) The Committee believes that two sets of obligations flow from these considerations. The first set relates to the affected State. The imposition of sanctions does not in any way nullify or diminish the relevant obligations of that State party. (…) While sanctions will inevitably diminish the capacity of the affected State to fund or support some of the necessary measures, the State remains under an obligation to ensure the absence of discrimination in relation to the enjoyment of these rights, and to take all possible measures, including negotiations with other States and the international community, to reduce to a minimum the negative impact upon the rights of vulnerable groups within the society. The second set of obligations relates to the party or parties responsible for the imposition, maintenance or implementation of the sanctions, whether it be the international community, an international or regional organization, or a State or group of States (CESCR 1997).

4 Policy Recommendations

Economic sanctions which produce significant negative impacts on population health are incompatible with the human right to health, which requires states to cooperate in order to ensure that every human being can achieve the highest attainable standard of physical and mental health. Economic sanctions are also incompatible with the principle of medical neutrality if they compromise physicians' capabilities to deliver healthcare in times of armed conflict or other humanitarian crises.

Table 12.4 provides policy recommendations for states when designing and implementing economic sanctions policies. In this context, it is important to emphasise that human rights and medical ethics should not only guide the design of economic sanctions policies, but also their implementation at all stages. Human rights and ethics reviews should be conducted not only before, but also during and after the implementation of economic sanctions. This review should take place at all

Table 12.4 Policy recommendations for economic sanctions to comply with the human right to health and the ethical principle of medical impartiality

Designing economic sanctions policies	Analysis of the (potential) health impact of the sanctions regime and its (potential) impact on vulnerable groups in the sanctioned country, e.g., children, persons with disabilities, persons living with chronic diseases; Comprehensive exemptions for the medical, pharmaceutical and biotechnology industry as well as humanitarian aid (and related sectors, e.g., banking & finance operations); Stakeholder consultation, e.g., with children's rights or humanitarian aid organizations
Implementing economic sanctions policies	Continuous assessment of the health impact of economic sanctions, particularly with regards to vulnerable populations, and of the impact of sanctions on global health and disaster & pandemic preparedness; Institution of independent (quasi-judicial) review mechanisms; Responsibility to **remedy** collateral impact of economic sanctions
Reviewing economic sanctions policies	Review of the health impact of current and finished sanctions regimes, adaption of current sanctions regimes to the findings; Compensation for unlawful measures (e.g., sanctioning measures incompatible with the human right to health)

levels of policy action on economic sanctions (national, regional and global), and should guide all sanctions policies. Sanctions are a key instrument to enforce respect for international human rights and humanitarian law, but in order to fulfil this mission, they themselves need to be designed and implemented in line with human rights and medical ethics (see also Germani et al. 2022).

5 Conclusion

Already back in 2000, then-United Nations Secretary General Kofi Annan has alerted that "just as we recognize the importance of sanctions as a way of compelling compliance with the will of the international community, we also recognize that sanctions remain a blunt instrument, which hurt large numbers of people who are not their primary targets" (UN 2000).

The present article has illustrated potential harms to (population) health which can be brought about by economic sanctions. In order to remain ethically acceptable instruments of international policy, economic sanctions must comply with the fundamental principles set forth by Article 1 of the Universal Declaration of Human Rights: "All human beings are born free and equal in dignity and rights. They are endowed with reason and conscience and should act towards one another in a spirit of brotherhood".

Even in times of conflict, it remains the responsibility of all states—including sanctioning and sanctioned states alike—to guarantee access of all humans to adequate health services and to ensure the respect of human rights and humanitarian law.

References

Aldebron, Jillian. 1993. Haitian AIDS education campaign struggles on despite political turmoil. *Aidscaptions* 1: 22–24.

Aloosh, Mehdi. 2018. How economic sanctions compromise cancer care in Iran. *The Lancet. Oncology* 19 (7): e334. https://doi.org/10.1016/S1470-2045(18)30427-3.

Aloosh, Mehdi, and Arash Aloosh. 2023. Economic sanctions and spread of infectious diseases. *Health Policy* 138: 104921. https://doi.org/10.1016/j.healthpol.2023.104921.

Aloosh, M., A. Salavati, and A. Aloosh. 2019. Economic sanctions threaten population health: The case of Iran. *Public Health* 169: 10–13. https://doi.org/10.1016/j.puhe.2019.01.006.

Asadi-Pooya, Ali A., Majid Nazari, and Nafiseh Mirzaei Damabi. 2022. Effects of the international economic sanctions on access to medicine of the Iranian people: A systematic review. *Journal of Clinical Pharmacy and Therapeutics* 47 (12): 1945–1951. https://doi.org/10.1111/jcpt.13813.

Barry, M. 2000. Effect of the U.S. embargo and economic decline on health in Cuba. *Annals of Internal Medicine* 132 (2): 151–154. https://doi.org/10.7326/0003-4819-132-2-200001180-00010.

Bezuidenhout, Louise, Ola Karrar, Javier Lezaun, and Andy Nobes. 2019. Economic sanctions and academia: Overlooked impact and long-term consequences. *PLoS One* 14 (10): e0222669. https://doi.org/10.1371/journal.pone.0222669.

Butler, Declan. 2019. How US sanctions are crippling science in Iran. *Nature* 574 (7776): 13–14. https://doi.org/10.1038/d41586-019-02795-y.

Carruth, Lauren. 2020. Medical neutrality. In *Humanitarianism: Keywords*, ed. Antonio de Lauri, 128–130. BRILL.

Council on Foreign Relations. 2022. *What to know about sanctions on North Korea*. https://www.cfr.org/backgrounder/north-korea-sanctions-un-nuclear-weapons.

Deilamizade, Abbas, and Sara Esmizade. 2015. Economic sanctions against Iran, and drug use in Tehran, Iran: A 2013 pilot study. *Substance Use & Misuse* 50 (7): 859–868. https://doi.org/10.3109/10826084.2015.978673.

Duric, P., and C. Peek-Asa. 2008. Economic sanctions, military activity, and road traffic crashes in Vojvodina, Serbia. *Injury Prevention : Journal of the International Society for Child and Adolescent Injury Prevention* 14 (6): 372–376. https://doi.org/10.1136/ip.2008.019240.

Financial Times. 2022. *African union warns of 'collateral impact' as EU's Russia sanctions hit food supplies*. https://www.ft.com/content/e558de33-6064-4b10-a784-eb344cb17915.

Germani, Federico, Julian W. März, Caroline Clarinval, and Nikola Biller-Andorno. 2022. Economic sanctions, healthcare and the right to health. *BMJ Global Health* 7 (7): e009486. https://doi.org/10.1136/bmjgh-2022-009486.

Gibbons, E., and R. Garfield. 1999. The impact of economic sanctions on health and human rights in Haiti, 1991-1994. *American Journal of Public Health* 89 (10): 1499–1504. https://doi.org/10.2105/ajph.89.10.1499.

Kim, Yiyeon. 2019a. Economic sanctions and child HIV. *The International Journal of Health Planning and Management* 34 (2): 693–700. https://doi.org/10.1002/hpm.2727.

———. 2019b. Economic sanctions and HIV/AIDS in women. *Journal of Public Health Policy* 40 (3): 351–366. https://doi.org/10.1057/s41271-019-00173-6.

Murphy, Adrianna, Zhaleh Abdi, Iraj Harirchi, Martin McKee, and Elham Ahmadnezhad. 2020. Economic sanctions and Iran's capacity to respond to COVID-19. *The Lancet. Public health* 5 (5): e254. https://doi.org/10.1016/S2468-2667(20)30083-9.

Pintor, Pinna, Marc Suhrcke Matteo, and Christoph Hamelmann. 2023. The impact of economic sanctions on health and health systems in low-income and middle-income countries: A systematic review and narrative synthesis. *BMJ Global Health, 8*(2). https://doi.org/10.1136/bmjgh-2022-010968.

Shahabi, Shohreh, Hooman Fazlalizadeh, Jennifer Stedman, Linus Chuang, Ahmad Shariftabrizi, and Regina Ram. 2015. The impact of international economic sanctions on Iranian cancer healthcare. *Health Policy* 119 (10): 1309–1318. https://doi.org/10.1016/j.healthpol.2015.08.012.

United Nations. 2000. Secretary-general reviews lessons learned during 'sanctions decade' in remarks to international peace academy seminar: Press release SG/SM/7360. https://press.un.org/en/2000/20000417.sgsm7360.doc.html.

———. 2022a. "Concerned by unintended negative impact of sanctions, speakers in security council urge action to better protect civilians, ensure humanitarian needs are met: SC/14788." https://press.un.org/en/2022/sc14788.doc.htm.

———. 2022b. Security council establishes sanctions regime on Haiti, unanimously adopting resolution 2653 (2022): SC/15073. https://press.un.org/en/2022/sc15073.doc.htm.

———. 2023. UN sanctions app. https://unsanctionsapp.com/.

United Nations Children's Fund. 2022. Sanctions and their impact on children: Discussion paper. https://www.unicef.org/globalinsight/media/2531/file/%20UNICEF-Global-Insight-Sanctions-and-Children-2022.pdf.

United Nations Committee on Economic, Social and Cultural Rights. 1997. General Comment 8, the Relationship Between Economic Sanctions and Respect for Economic, Social and Cultural Rights: U.N. Doc. E/C.12/1997/8. http://hrlibrary.umn.edu/gencomm/escgencom8.htm.

———. 2000. General Comment No. 14: The Right to the Highest Attainable Standard of Health: U.N. Doc. E/C.12/2005/4. https://www.ohchr.org/en/resources/educators/human-rights-education-training/e-general-comment-no-14-right-highest-attainable-standard-health-article-12-2000.

United Nations General Assembly. 1948. Universal declaration of human rights: U.N. Doc. A/RES/217(III). https://www.un.org/en/about-us/universal-declaration-of-human-rights.

———. 2002. Resolution: elimination of unilateral extraterritorial coercive economic measures as a means of political and economic compulsion: A/RES/57/5. https://digitallibrary.un.org/record/476325?ln=en.

United Nations Human Rights Council. 2014. Human rights and unilateral coercive measures: A/HRC/RES/27/21. https://ap.ohchr.org/documents/dpage_e.aspx?si=A/HRC/RES/27/21.

United Nations Security Council. 2023. Sanctions – Fact sheet. https://www.un.org/securitycouncil/sanctions/information.

World Health Organization. 1946. Constitution of the WHO. https://www.who.int/about/governance/constitution.

World Medical Association. 2022. WMA declaration on the protection and integrity of medical personnel in armed conflicts and other situations of violence: Adopted by the 62nd WMA general assembly, Montevideo, Uruguay, October 2011 and Revised by the 73rd WMA General Assembly, Berlin, Germany, October 2022. https://www.wma.net/policies-post/wma-statement-on-the-protection-and-integrity-of-medical-personnel-in-armed-conflicts-and-other-situations-of-violence/.

Part IV
Challenging Medical Neutrality

Chapter 13
A Practical Reflection on Global Health Leveraging Health as a Means to Another End

Melissa McRae

Abstract Our understanding of health has been expanding for decades but was boosted by COVID-19. The list of variables influencing health feels infinite, and what constitutes an actual health intervention by health professionals is no longer implicit. This paper explores if the discipline of global health is furthering the expansion of health and what that means for patients and health care providers who implement global health initiatives. Using the WHO Global Health for Peace Initiative (GHPI) as one example, the author explores if there is an increasing tendency to leverage healthcare for non-direct health outcomes. A particular focus is given to implementation impact for humanitarian populations if the medical mission is transformed from being an 'end goal' to a 'means to another end'. Médecins Sans Frontières (MSF) has been analysing the potential operational evolution and ethical implications of the GHPI. In this analysis the author brings together views on how global health is framed in the literature and why it is currently so popular as a specific field, followed by an analysis of three potential ethical risks at the implementation level: (i) health as a means to another end, (ii) upholding medical neutrality, and (iii) supporting the health workforce.

Disclaimer: The views expressed in this chapter are personal and may not reflect the position of Médecins Sans Frontières. Additionally, they reflect on the current formulation of the GHPI initiative at the time of writing. The research and writing occurred early in 2023 as the GPHI was undergoing first and second round consultation.

M. McRae (✉)
Médecins Sans Frontières, Geneva, Switzerland
e-mail: melissa.mcrae@geneva.msf.org

D. Messelken, A. E. Barbar (eds.), *Challenging Medical Neutrality*, Military and Humanitarian Health Ethics, https://doi.org/10.1007/978-3-031-69398-4_13

1 Introduction

The COVID-19 pandemic took a catastrophic toll on health, lowering global life expectancy at birth for the first time since the 1960s (Pirlea and Suzuki 2023; World Bank 2023). Declared as a global health emergency, it reaffirmed that health is universally morally important and demonstrated the co-dependencies between individual health, health systems and how societies govern access to healthcare. Health clearly emerged from COVID-19 a political asset but also a moral good to be protected from politicisation (Sauter 2023; van Stekelenburg et al. 2023). Nations were affected and responded unequally, highlighting gross global health disparities and inter-variability in public health approaches. All this boosted support for the discipline of global health which promotes what public health specialists have been telling us for years: achieving optimal health necessitates more than *just* a healthcare response.

Global health places a *"priority on improving health and achieving health equity for all people worldwide"* (Koplan et al. 2009), promoting health inequities as morally problematic realities that persist due to structural, socially constructed political barriers. Since its emergence in the early 2000s, global health has scaled up from origins in epidemic disease control to now span a plethora of topics: health system strengthening, outbreak management, health security, border security, international relations, arts, and health, and recently the World Health Organisation (WHO) Global Health and Peace Initiative (GHPI) which has healthcare actors contributing to peace building and sustainability.

As explored in this paper, among the Pandora's box of global health initiatives, health is widely interpreted allowing numerous endpoints; many with direct health benefits, others with non-direct health benefits and for some, the link to health benefits can be tenuous. Global health has the difficult task of generating political and socioeconomic support to address inequities without slipping into the dangerous space of politicising health, a *"process, whereby political discourse and rhetoric seep into the legislation and governance of healthcare, [negatively impacting] its functioning and efficiency"* (Sibanda 2019). This is a terrain well-suited to ethical deliberation, and yet global health ethical guidance has been found to be underdeveloped and an afterthought (Robson et al. 2019).

Historically, understanding health to be more than the absence of disease can be traced back to 1948 when the WHO reformed the definition of health to be a *"state of complete physical, mental and social wellbeing"* (WHO 1948): pushing traditional medicine to recognise mental and preventative health needs, and the social determinants of health (poverty, education, housing, employment) as crucial for wellbeing. This act reframed a public health orientated view of health as a *means* rather than an endpoint, a shift at odds for those who *provide* care to act in the best interest of patients' and populations' health as their end goal, the concept of health primacy (Galea and Vaughan 2019). Thinking beyond the well-known ethical tensions between preventative and curative health, conceptualising health as a means through which *"individuals can live full and rich lives"* highlights just how many influential components lie outside the ethical remit of the health profession, also referred to as non-health dimensions of well-being (Galea and Vaughan 2019; Merritt et al. 2018). Yet, despite this expansion, I suspect many people still associate the word health with the work of health professionals. How is global health balancing the new space for these dimensions without creating confusion at the implementation level over the roles and responsibilities of the healthcare profession and patients' access to healthcare?

In this expanding domain of global health, what constitutes a health intervention delivered by healthcare providers (HCPs) might no longer be implicit for patients, the public, nor those within the profession. This opens the door for the moral value of health to be leveraged for secondary agendas that may, or may not, improve well-being but do not bring direct, tangible health benefits to patients and populations, all in the name of health. For years health professionals have questioned the utility of the WHO definition, referring to it as *"unhelpful"* and even *"counterproductive"* (Godlee 2011; Smith 2008). Firstly, it dismisses the positive attributes of human capabilities to cope and be well whilst living with ageing, chronic illness and disability (Godlee 2011). Secondly, complete health is likely unachievable for most of us given biological imperfections but, more importantly, is complete health morally what we should be striving for (Smith 2008)? Health is shifted into an unmeasurable, unreportable state, contributing to unclear, maybe unrealistic, expectations of healthcare practices, subsequently threatening the trusting relationship on which the health profession relies (Godlee 2011; Smith 2008). This trend did not begin with the field of global health, but this paper explores if it is being propelled by it and without ethical guidance.

Trust is required for HCPs to deliver care in all contexts, but when providing humanitarian response in armed conflict settings a loss of trust can have very severe consequences. In the latter, it is the medical mission that fosters trust and secures access to healthcare for those in need. This mission requires a clearly shared understanding by all parties of the role of HCPs and their exclusive assignment to medical duties: providing care being their first and only professional duty (Red Cross 2011). This is a caveat for upholding medical neutrality, an ethical principle and legal construct to provide protective immunity from harm for patients and staff working in armed conflict where attacks on health care still persist (WHO 2023d). These are the aspects to be weighed against the added value of health being so broadly interpreted and politically supported may bring globally.

Médecins Sans Frontières (MSF) has been analysing the potential operational evolution and ethical implications of the GHPI as one example of a global health initiative that, from a humanitarian perspective, requires ethical reflection before implementation in conflict settings (MSF 2023b). In this paper the author explores whether health is becoming so poorly defined that some global health initiatives are leveraging health 'as a means to another end'. The analysis draws on available literature (published and grey) and the author's experience in humanitarian contexts providing a practical probing of the potential implications for patients and HCPs if the medical ethical obligation to prioritise a patient's health as the primary endpoint is compromised (WMA 2021). Section one brings together a summary of the vast terrain global health is attributed to in the literature and explores three pillars that contribute to its popularity as a specific field: partnerships, health equity and health politics. Section two provides an outline of the Global Health and Peace Initiative in 2023, and section three discusses from a humanitarian perspective the potential impact of three normative and operational risks, of GHPI implementation in conflict settings: (i) health as a means to another end (ii) upholding medical neutrality, and (iii) supporting the health workforce. To be clear, this case study is dissecting just one of many global health initiatives, but one with potential operational implications which MSF has spent some time trying to better understand (MSF 2023a, b). It is written from the author's perspective as a humanitarian worker trying to better articulate potential opportunity costs related to well-intentioned global health initiatives in conflict affected humanitarian contexts.

2 What Is Global Health?

It seems global health is frequently referenced but rarely defined, becoming an umbrella term covering a multitude of initiatives from the academic theories of human rights, medicine, natural science and social science (Beaglehole and Bonita 2010; Hall-Clifford and Cook-Deegan 2019b; Koplan et al. 2009). Global health aims to improve coordination within the sector whilst also cross fertilising health with traditionally non-health-related political sectors (employment, economics, finance) to better address the complexity of societal transactions required for healthy populations, often governed by politically influenced systems (FDFA 2014; Koplan et al. 2009; Lincoln 2021). Despite academic efforts, no universal definition exists, permitting enormous plasticity for global health to be framed to meet a myriad of different objectives (Abimbola 2016; King et al. 2019; Koplan et al. 2009; Lincoln 2021; Robson et al. 2019). As the normative global health body, WHO directs global health initiatives to *"tackle increasing global health threats, reduce disparities within communities and between nations and contribute to a world where people live healthier, safer and longer lives"* (WHO 2023c) Many WHO supported initiatives exist,[1] ranging from arts to climate change, whilst beyond WHO, global health is also viewed to tackle pharmaceutical evergreening, TRIPS Plus, the opioid epidemic, resource conflicts, healthcare infrastructure rebuilding in fragile & post-conflict states, rape as a weapon of war, drone warfare, or the global refugee surge (DeCamp et al. 2018). The scope seems boundless, embracing any efforts that make our lives happier, safer, and longer as a health initiative!

[1] Access to COVID-19 Tools (ACT) Accelerator, Alliance for the Global Elimination of Trachoma by 2020 (GET 2020), Behavioural Sciences for Better Health, Alliance for Transformative Action on Climate and Health (ATACH), Arts and Health, Global Health, and Peace Initiative (previously Global Health for Peace), Advanced global health security: from commitments to actions, Action on Social Determinants of Health for Advancing Health Equity.

Global health belongs to several disciplines: academic study, research and practice (Koplan et al. 2009). Alternatively, Peter Farmer et al. (2013) view global health to be "*not [an academic] discipline but a collection of problems [that] ...turn on the quest for equity*". In 2022, the search engine PubMed cited more than 21,846 articles containing the term "global health", up from 10,255 in 2015[2] (PubMed 2023). In a systematic literature review searching for "global health" AND ethic*" between 1977 and 2015, the earliest articles dated back to the early 2000s and publications peaked in 2011 (29 articles) (Robson et al. 2019). The authors revealed two important but unlinked findings, firstly the 'global South' was underrepresented, with 88 per cent of articles by authors from high-income countries and secondly, the dimension of ethical guidance for global health *"incompletely understood or developed"* (Robson et al. 2019). The latter is supported by other authors who attributed this paucity of ethical guidance to the wide scope of global health themes trying to be addressed under one umbrella such that existing bioethical standards fall short (Godard et al. 2018; Hall-Clifford and Cook-Deegan 2019a).

The nomenclature 'global health' hails from North American, developed to complement the practice of International Health from the 1980s and 1990s (Koplan et al. 2009). International health had origins closer to Europe, amalgamating tropical medicine (infectious diseases) and public health, with a particular focus on maternal and child health. The latter is critiqued to address health issues stemming from institutions in high-income countries whilst relating almost exclusively to work abroad in low-resource nations (Beaglehole and Bonita 2010; Koplan et al. 2009). Global health has expanded the terrain; for example the Swiss government described international health to have "*focused primarily on bilateral relations between high-income and low-income countries [...] whilst global health places greater emphasis on the multiplicity of relationships required, on a global scale*" (FDFA 2014). This evolution likely serves the purpose of ensuring the words we use reflect changes in values—actual or desired—such as how countries cooperate and navigate power dynamics, social relations, inequalities, and politics (Table 13.1).

In attempting to understand the vast prevue of global health, authors have queried both 'what terrain global health covers' and 'what is new via global health'. As an overarching term, global health nearly always refers to the worldwide span of who is affected by a problem, as evidence from the UK government's global health definition: *"where the determinants [...] are beyond the capacity of individual countries to address through domestic institutions"* (Public Health England 2014). Global health also embraces all three health system levels: the micro (individual, clinical), meso (regional, institutional) and macro (national and provincial), compared to public health prioritising the latter two. As explored below, it is an ambitious task to balance the old ethical tensions of public and individual health, whilst also explicitly raising health's political profile sufficiently to address structural barriers to health equity. Despite the diversity of initiatives, some common threads and

[2] Search last verified 04 August 2023.

Table 13.1 Comparison of global, international, and public health

	Global health	International health	Public health
Geographical reach	Focuses on issues that directly or indirectly affect health but that can transcend national boundaries	Focuses on health issues of countries other than one's own, especially those of low-income and middle-income	Focuses on issues that affect the health of the population of a particular community or country
Level of cooperation	Development and implementation of solutions often requires global cooperation	Development and implementation of solutions usually requires binational cooperation	Development and implementation of solutions does not usually require global cooperation
Individuals or populations	Embraces both prevention in populations and clinical care of individuals	Embraces both prevention in populations and clinical care of individuals	Mainly focused on prevention programmes for populations
Access to health	Health equity among nations and for all people is a major objective	Seeks to help people of other nations	Health equity within a nation or community is a major objective
Ranges of disciplines	Highly interdisciplinary and multidisciplinary within and beyond health sciences	Embraces a few disciplines but has not emphasised multidisciplinarity	Encourages multidisciplinary approaches, particularly within health sciences and with social sciences

Source: Koplan et al. (2009, p. 1994)

recurring pillars of global health are explored below; partnerships, addressing health inequities and health politics.

2.1 *Partnerships*

Two-way collaboration and cooperation via multisectoral partnerships (local and international) is a flagship pillar of global health, an innovative step to welcome non-traditional actors into the health sector and an attempt to rebalance North-South dynamics (FDFA 2014; Koplan et al. 2009; Public Health England 2014). A focus on subsidiarity, reaffirms that local and proximate collaboration is essential to achieve health equity and positively challenges previous North-South power imbalances that plagued international health from colonial days (Abimbola 2021). However, diverse partnerships, particularly if formed without common ethical values, also creates space for new power dynamics to emerge and introduce a potential lack of transparency of intentions and agendas (Lincoln 2021). For example, global health has been described as an avenue for "*some of the world's most powerful political and economic institutions to advance their [own] interests*" (Lincoln 2021). As explored in Section three, in humanitarian contexts, a proliferation of actors with mandates beyond health services may divert attention from the medical mission and concerningly lower the value of lifesaving action where direct healthcare remains the largest need.

Governments, military forces, universities, private sector philanthropists, as well as non-governmental humanitarian organisations, such as MSF, are all embracing global health initiatives (King et al. 2019; Lim and Blazes 2015). Research shows that global health partnerships exist along a spectrum, with political economic partnerships at one end and interpersonal (HCP-patient) relationships at the other; rarely are both found in a single initiative (Robson et al. 2019). This may demonstrate that global health ideologically, holistically spanning of all system levels is not feasible in practice. This will raise problems for successful implementation of global health initiatives, but coincidently may provide a protective separation when politically supported health systems and governance must be differentiated from individual care that must remain impartial and non-politically influenced to prioritise patient needs and protect health primacy.

Whilst a government health system focus is crucial in addressing equitable access to care, in humanitarian contexts the same degree of health system focus can be problematic for minorities and marginalised populations whose needs maybe justifiably excluded via the utilitarianism ‘greater good’ logic adopted by many government health systems. Health systems in humanitarian contexts can be fragile, impeded from optimal performance, and political support is either not readily available or health is being politicised to exclude such populations from accessing healthcare. Section three provides examples of operational risks related to global health confabulating these two ends of the spectrum.

2.2 Addressing Health Inequities

Global health promotes the popular view that all global citizens are equally deserving of the opportunity to be healthy. Striving to address inequities everywhere—between and within low-, middle- and high-income countries—is an aspect that Professor Abimbola (Editor-in-Chief, BMJ Global Health Journal) believes is the salient distinction of global health (Abimbola 2016). Yet of course, before global health emerged, for decades public health has been addressing the social determinants of health, *“the causes of the causes”* to quote Sir Michael Marmot. Admittedly, these efforts are said to have been plagued with implementation challenges attributed to the chronic underdevelopment of the politics of health as expanded below (Kusnanto et al. 2018; Marmot 2018).

Addressing inequities is one reason for the popularity of global health as Nation States can simultaneously uphold moral obligations owed to less privileged nations, whilst also protecting the health of their own citizens, heeding the warning of WHO Director-General Dr. Tedros Ghebreyesus (2020) that *“we are only as safe as the weakest link”*. This aspect firmly cements global health as a pillar of foreign policy, framed under another new term: Global Health Diplomacy (GHD). GHD is *“the process by which government, multilateral and civil society actors attempt to position health in negotiations foreign policy and to create new forms of global health governance"* to reduce poverty and inequalities. GHD is said to have created high expectations and many collaborations but only so far amounted to good intentions and not delivered a positive change in the relations between high- and low-income countries (Rubbini 2018). Additionally, this aspect is linked to another broadly interpreted concept of global health security, which is beyond the scope of this paper (FDFA 2014; Lim and Blazes 2015; Rubbini 2018).

2.3 Health and Politics

So, in this quest for equity, global health is explicitly seeking political partnerships, financial and socioeconomic, to abolish the ideological belief that health is apolitical (Bambra et al. 2005). Yet WHO's definition of health is critiqued for promoting this belief, by solely focusing on individuals, not societies. Health then becomes a product of individual factors with the flow on effect that health is incorrectly reduced to healthcare (Bambra et al. 2005). Bambra et al. assert that health *is* political, as the determinants of health are amenable to political action and influenced by political ideology of how societies should co-exist; yet the politicisation occurs when the political agenda influences how healthcare is accessed by patients and delivered by HCPs (Bambra et al. 2005).

The politicisation of health is said to have effectively started in Western countries in the 1970s. With increasing public spending going towards individual curative care to address preventable illness (diabetes, obesity, accidents), politicians took note of the cost efficiency promoted via public health interventions (Sibanda 2019). Two consequences arose: firstly, politicians began supporting national public health policies but secondly, health became politicised by the reinforcing of privatised healthcare. Health was elevated to a super value market commodity for affluent individuals; money meant access to healthcare, money represented all that is good in life (Crawford 1980; Sibanda 2019). Of course, for the non-affluent, health inequity gaps widened. However it was not an expansive re-definition of health that made politicisation inevitable but the large-scale financial involvement of government in individual care (Sibanda 2019). This economic pragmatism also influences global health, introducing a reality check between global health aspirations and what can be concretely implemented in practice (Rubbini 2018).

Calls urging for high-level political support have been made by MSF, among others, often during international health crises such as the HIV-AIDS pandemic, West African Ebola epidemic, and of course the COVID-19 pandemic (MSF 2002, 2014). All highly political public health events requiring political solutions which demonstrate the ways in which power, political and ideology influence health (Bambra et al. 2005). Yet the political nature of health is infrequently featured in global health discourse (Gore and Parker 2019). As explored in this case study, ethical guidance is needed to safeguard health's political profile for optimal societal benefit, whilst simultaneously protecting the professional ethics of the medical mission from secondary agendas. If successful, much needed political support to health systems and policies will emerge. Alternatively, and likely if ethical guidance is lacking, politicisation will expose the conflicts that exist in different system layers, between policymakers who follow government agendas and healthcare professionals and patients orientated to health primacy (Ball and Corbis 2015; Sibanda 2019). At the implementation level, this politicisation has been shown to be most detrimental for patients and HCPs, not policy makers nor health management (Sibanda 2019). Putting the two in competition may be an ethical reality but is not a productive one.

> On some issues, ... the position of physicians may have to be promoted forcefully against those of governments, health system administrators and/or commercial enterprises. World Medical Association Medical Ethics Manual. (Ball and Corbis 2015, p28)

The medical mission and delivery of care to individuals must remain a non-political act, especially in conflict zones, where any misinterpretation of political alliances can be fatal (WHO 2023d). Despite this understanding, unfortunately politically motivated attacks on healthcare happen everywhere (Fast and Roborgh 2020). COVID-19 research demonstrated that HCPs are frequently, yet incorrectly, perceived to be accountable for politically orientated health decisions and that inaccurate government information plays a role in citizen distrust towards HCPs (van Stekelenburg et al. 2023). This leads to different forms of aggression experienced in their professional and personal lives, partially attributed to the politicisation of the COVID-19 response (Fast and Roborgh 2020; van Stekelenburg et al. 2023).

Finally, global health has not come from nowhere, it is a product of evolving past practices, theories, and perceptions of how we view health and healthcare in today's world, yet the daily implementation of healthcare, delivered by HCPs, is not well differentiated from health as a political contrast. Whilst the agendas of global health are positive and aspirational, the implementation impact calls for "*substantive reconceptualisation of the political, ideological and material foundations of global health interventions*" (Lincoln 2021). Examples given include global health practices which led to the withholding of clinical resources in crisis settings, removal of public health assistance when programme targets failed to be met and concerningly, the planned commercialisation of medicine amid economic collapse (Lincoln 2021). The next section analyses some other potential implementation risks.

3 A Global Health Case Study: Global Health and Peace Initiative

In the words of Dr. Tedros Ghebreyesus, "*there cannot be health without peace, and there cannot be peace without health*" (WHO 2021). The merit of this tenet highlights the importance of contextualising the following analysis specifically to **humanitarian response in conflict settings** (WHO 2022). As "*enhancing the* *existing links between health and peace"* is the driving force behind a relatively new WHO initiative, sponsored by the States of Switzerland and Oman, launched in 2019 as 'Global Health for Peace Initiative' (GHPI) and brought for consultation with Member States and WHO Non-State actors in Q4 2022 and Q1 2023 (WHO, 2023b).

As a collaboration between WHO and the peacebuilding organisation Interpeace, the GHPI aims to *"position health as an influencer of peace"* and the WHO as a sustaining peace actor: strengthening WHO's role, and by affiliation the health sector's, as contributors to equitable healthcare access, health system strengthening, the expansion of universal healthcare and contributing to the Triple Billion targets (WHO 2023a, b). WHO is seeking a means for health actors to be better able to address the underlying drivers of critical health needs in *"fragile, conflict-affected, and vulnerable settings"*. Settings broadly framed by WHO to include humanitarian crises, protracted emergencies, and armed conflicts, where peace is desperately needed. Approximately 80% of WHO's humanitarian caseload and 70% of the disease outbreaks to which they respond are found in these settings (WHO 2023).

Early drafts of the GHPI Roadmap (September 2022) primarily frame the initiative to contribute to creating and sustaining peaceful societies, referring *"to 'small p' peace, such as social cohesion, trust, inclusion, resilience to violence (rather than 'big P' Peace, in the form of high-level political processes/solutions)"* (WHO 2023). First and foremost, defining peace in order to obtain a common understanding during the consultation period proved challenging. Additionally in September 2022, GHPI described health actors transitioning from working 'in' conflict to working 'on' conflict, building on previous WHO programming known as 'Health as a Bridge for Peace' (1980s and 1990s) (Schneider 1987; WHO 2022). This change of "on" conflict aim to incorporate delivering health programmes, where feasible and appropriate, integrated with peacebuilding strategies; *"deliver peace dividends through health interventions*" (WHO 2022).

The GHPI Roadmap include six workstreams and four principles (WHO 2023). The principles are context specificity, equity and inclusiveness, participation and local ownership and leadership. The six workstreams include: (1) Evidence generation through research and analysis, (2) Development of a strategic framework, (3) Advocacy and awareness-raising, (4) Capacity-building, (5) Mainstreaming of the Health for Peace approach and (6) Partnership development. MSF provided feedback that considerations of individual needs (micro layer) initially appeared scant, with little attention to patient factors: standards of care, quality of care or ethics of any kind– neither medical, public health or humanitarian (MSF 2023b). Through consultation the GHPI Roadmap draft (January 2023) included medical ethics and humanitarian principles as a fifth principle.

The GHPI defines its target audience as 'health actors', implicitly understood to refer to the broad, heterogenous group of professionals working across all three layers of the health system. Clarification was sought on how WHO differentiated various professional health roles and responsibilities, for example those working on health system strengthening and programming from those delivering care at the bedside. MSF's concern was for HCPs whose obligations to patients might raise ethical concerns, real or perceived, through involvement in delivering peace dividends, compromising medical neutrality and potentially the medical mission. In later drafts, it became evident but still not explicit, that the GHPI is fundamentally targeting a political public health audience, Member States' Ministries of Health (MoH) and the WHO Secretariat, roles more aligned with the political aspect of health at a programming and policy level. The current Roadmap states that healthcare workers at the implementation level are *"not expected to take on responsibilities outside their existing medical mandate and GHPI programming will always comply with medical ethics"* . This version was produced for the World Health Assembly (WHA) in May 2023 and introduced a significant change of name, from Global Health *for* Peace to Global Health *and* Peace (WHO 2023).

4 A Humanitarian Conflict Setting Orientated Analysis

Through work done with MSF, the author is still trying to understand the GHPI in concrete operational terms. It is critical to comprehend the actual ask of MSF HCPs and the government appointed HCPs MSF staff work alongside through MSF-MoH partnerships in more than 60% of MSF Switzerland operations.[3] This level of partnership brings MSF operations in conflict settings into close contact with global health initiatives supported by WHO Member States.

Without doubt, the ideology behind this initiative comes with good intentions, peace is essential for healthy living, but without a better understanding of the implementation approach it is prudent to stay alert to potential opportunities and risks explored below. This analysis focuses on the impact for patients seeking MSF care and MSF staff if healthcare interventions are used, or perceived to be used, to deliver peace dividends in conflict zones. Recognising this is done pre-emptively as the GHPI Roadmap is still in consultation phase although the potential impact of the GHPI could be significant and wide-reaching given the influential normative role of WHO policies across the entire global workforce.

4.1 Normative Risk: Leveraging Health Interventions for a Means to Another End

> If, in performing their professional duty, physicians have conflicting loyalties, their primary obligation is to their patients; Privileges and facilities afforded to [...] healthcare professionals [...] must never be used other than for healthcare purposes. (WMA 2022b)

[3] MSF Switzerland internal data. As an international movement MSF is operational in 72 countries in 2023. MSF Switzerland is operating 58 projects in 27 countries. A total of 22 (38%) are in armed conflict or unstable contexts. Of those, 80% are run in partnership with the local MoH.

This ethical guidance from the World Medical Association (WMA) is clear and echoes Kantian moral philosophy: to *"act in such a way that you treat humanity [...] never merely as a means to an end, but always at the same time as an end"* (Kant and Reath 1997). A HCP must prioritise patients' health as the primary endpoint above other competing interests (WMA 2021). This ethical norm safeguards a transparent and trustworthy person-orientated relationship between provider and patient, which in part safeguards access to care.

By proposing to *"deliver peace dividends through health interventions"*, healthcare becomes a means to achieve another end (WHO 2022). The two may not be mutually exclusive endpoints but health primacy is at risk of compromise. If addressing direct health needs is no longer the primary objective of a healthcare intervention led by HCP, then the profession is entering into ethically dangerous waters. The repurposing of health interventions as a means to another end is an old, not new, lesson global health must adhere to. Past examples are numerous (syphilis study in Tuskgekee 1932, medical experiments to the Nuremberg Code 1947), both foundational in creating the discipline of bioethics HCPs must follow today (British Medical Journal Publishing Group 1996; Butler 1973). Another more recent example is the fake vaccination campaign which took place in Pakistan in 2011, associated with the capture of Osama bin Laden (Kennedy 2017). Whilst an extreme example, it demonstrates that health being used for secondary agendas, particularly political ones, is not just a theoretical risk. More than a decade on, the consequence is reduced community trust towards HCPs, fueling vaccine hesitancy and resulting in a 23–39% decline in childhood immunisation and an increase in preventable disease prevalence (Martinez-Bravo and Stegmann 2022). When such health leveraging occurs, it is patients and populations who suffer the consequences.

HCPs also have an ethical obligation to ensure they *"avoid acting in such a way as to weaken public trust in the medical profession"* (WMA, 2022a). Leveraging health impacts how patients (and populations) both receive and perceive healthcare. Firstly, healthcare risks being disconnected from delivering direct health benefits and secondly, the roles and responsibilities of HCPs, who directly interact with patients, can be negatively perceived and reshaped. A HCP's duty to uphold non-maleficence and beneficence can be questioned, and transparency lost. This could ultimately diminish trust in the profession which impedes access to quality care. If this eventuates, it may also contribute to workforce implementation challenges if HCPs feel their professional ethical obligations are compromised. Acknowledging the GHPI does state health outcomes should remain a priority, there is a lack of ethical guidance to help enforce this endpoint and prevent a slippery slope towards leveraging and crucially, this leveraging being perceived to be done by HCPs themselves. A clear definition of what a health intervention is within the remit of the GHPI would also assist in formulating more concretely the implementation reality. Of note, this normative risk is, in the author's opinion, what prompted a change in the initiative title from Global Health *for* Peace to Global Health *and* Peace.

4.2 Operational Risk: WHO, Member States and All Parties to a Conflict Must Uphold, Protect, and Promote Medical Neutrality

The operational risk centers around HCPs being assigned additional non-healthcare duties in order to deliver peace dividends and work 'on' conflict. The principle of medical neutrality permits immunity from being a target in conflict *only* if HCPs exclusively deliver medical duties (Red Cross 2011). Medical neutrality means adhering to non-interference and exclusive assignment of HCPs to 'medical duties', any reassignment away from these duties, again real or perceived, compromises the specific protection given to HCPs under International Humanitarian Law (Red Cross 2011). From operational experience the author can attest that for those who implement a system, their lived experience acts as a mirror of the ethical status of that system. Perhaps more influential than law in some contexts, medical neutrality is also part of a social contract, contributing to social cohesion practices and peaceful coexistence in its own way. Hence, immunity from attack is granted via a 'convention of social agreement', in addition to being a moral imperative, and everyone pays the cost of breaching this convention (Gross 2007).

In humanitarian operations it is respect for medical neutrality that helps protect healthcare facilities as neutral, non-political safe havens where respect for human life and optimal health take priority. The GHPI fails to recognise that it is medical neutrality is what permits HCPs to be both beneficiaries and contributors to peace. When HCPs are no longer believed to be solely providing healthcare, political influences and suspicions of secondary agendas can very quickly emerge. Local rumours start to circulate and HCPs and patients can become direct targets, often executed via an attack on the healthcare facility they work in (Fast and Roborgh 2020). Whilst ideologically the relationship of health and peace may be symbiotic virtuous endpoints, in conflict zones, misinformation and disinformation are rampant, and transparency of action is foundational in promoting a safe working environment.

Furthermore, health providers must act in accordance with their professional competences and ethical code, transparently and consistently, to 'do no harm' in order to earn this trust from patients and the community. This is a fragile relationship, particularly in conflict zones. Trust in healthcare is an unwritten agreement, often between strangers, that is easily lost and hard to regain, both at the patient and community level, implicating individual HCPs and the organisations they represent. Whilst now removed from the language of the GHPI Roadmap draft, what triggered MSF's thinking on this normative risk was HCPs being asked to work 'on' conflict (MSF 2023b; WHO 2022). The fear was the introduction of a conflict of interest within the roles and responsibilities of HCPs. This fear is based on years of implementation experience where we have learnt that the perception of a conflict of interest can be just as damaging to a patient-provider relationship as a substantiated conflict of interest. Optimal healthcare gives patients no reason to suspect that a HCP is not acting in the best interest of their health at all times.

4.3 *Normative Risk: Inappropriate Allocation of Responsibility to Healthcare Providers, an Already Fragile Workforce*

A second normative risk is of vital importance given the existing baseline fragility of the healthcare workforce in many humanitarian, conflict affected settings. During initial GHPI consultations, in addition to seeking clarity of the perception of a HCPs roles and responsibilities as discussed above, MSF questioned why HCPs are being given such a large responsibility for "peace building" beyond their professional remit.

Contributing to peace is the responsibility of all global citizens, and HCPs should not be burdened with a higher moral responsibility than others to address the drivers of conflict and deliver peace dividends. The health workforce should not be inappropriately given responsibilities that do not fall within their professional scope of practice; working within one's professional competence is an ethical obligation (WMA 2022a).

A recent editorial in the Lancet Global Health titled 'Health-care workers must be trained and retained' highlighted that working as a HCP is already an extremely difficult role and a profession people are opting out of (The Lancet Global Health 2023). The WHO 'Health workforce support and safeguards list progress report' added eight countries with critically low workforce densities to the 2020 list of vulnerable countries (WHO 2023a). Of 55 identified countries, 26 are currently in active conflict, 65% of these being contexts where MSF has operations (MSF 2023a). Drawing lessons from the COVID-19 pandemic, HCPs are vulnerable to being overworked, to workplace violence, and are increasingly questioning demands from authorities and employers which compromise their professional ethical integrity (Fast and Roborgh 2020). Whilst the concrete activities linked to the GHPI for HCPs is still not clear, new global health initiatives should help, not exacerbate the already high burden on the health workforce, especially those who work in high-risk contexts such as conflict. As a risk this could negatively impact the physical and mental wellbeing of HCPs and their job satisfaction.

5 Conclusion

Global health is picking up speed and evolving in many directions with a firm moral intention of "*improving health and achieving health equity for all people worldwide*" (Koplan et al. 2009). Yet global health is now a widely shared space introducing diverse agendas from multiple partners, many of whom are explicitly trying to make inroads in health through optimising political support to address health inequities. Whilst still theoretical at the time of writing, this analysis of WHO's GHPI initiative highlights the need for stronger ethical deliberations on realistic risks that could eventuate during the implementation of global health initiatives; taking the GHPI implementation as one case in point but realising it is likely relevant to other initiatives.

Conflict undeniably costs human lives and health outcomes are categorically worse in conflict settings, So, whilst we do need initiatives to sustain peace they must be implemented to protect health primacy and reinforce a trusting patient-provide relationship. Equally, both health primacy and peace need to be protected, one should not come at the cost of the other. If via global health thinking, health is leveraged as a means to another endpoint—even if a very noble one like peace—we must simultaneously ensure that HCPs are supported in their roles and able to uphold their professional ethical obligation to patients.

In conclusion, an uncritical acceptance of the solid moral objectives of global health contributing to a greater social good, ignores the potential negative impact of the leveraging and politicisation of health for secondary agendas. This will be highly detrimental to the immediate health needs of people living in conflict zones, patients, and front-line HCPs. Global health ethics is an underdeveloped area and would benefit from more ethical rigor and nuance to not negatively impinge on patient's access to care and the trust and safety of the global health workforce.

References

Abimbola, S. 2016. The information problem in global health. *BMJ Global Health* 1 (1): e900001. https://doi.org/10.1136/bmjgh-2015-900001.

———. 2021. The uses of knowledge in global health. *BMJ Global Health* 6 (4): e005802. https://doi.org/10.1136/bmjgh-2021-005802.

Ball, R., and Corbis. 2015. *Medical ethics manual*. World Medical Association.

Bambra, C., D. Fox, and A. Scott-Samuel. 2005. Towards a politics of health. *Health Promotion International* 20 (2): 187–193. https://doi.org/10.1093/heapro/dah608.

Beaglehole, R., and R. Bonita. 2010. What is global health? *Global Health Action* 3. https://doi.org/10.3402/gha.v3i0.5142.

British Medical Journal Publishing Group. 1996. The Nuremberg code (1947). *BMJ* 313 (7070): 1448. https://doi.org/10.1136/bmj.313.7070.1448.

Butler, B.N. 1973. *Final report of the Tuskegee Syphilis Study Ad Hoc Advisory Panel (1973)*. US Department of Health, Educaiton and Welfare. https://biotech.law.lsu.edu/cphl/history/reports/tuskegee/complete%20report.pdf.

Crawford, R. 1980. Healthism and the medicalization of everyday life. *International Journal of Health Services* 10 (3): 365–388.

DeCamp, M., L.S. Lehmann, P. Jaeel, C. Horwitch, A.C.P. Ethics, and Professionalism and Human Rights Committee. 2018. Ethical obligations regarding short-term global health clinical experiences: An American College of Physicians Position Paper. *Annals of Internal Medicine* 168 (9): 651–657. https://doi.org/10.7326/M17-3361.

Fast, L., and S. Roborgh. 2020, April 28. Healthcare workers are still coming under attack during the coronavirus pandemic. *The Conversation*. http://theconversation.com/healthcare-workers-are-still-coming-under-attack-during-the-coronavirus-pandemic-136573.

FDFA. 2014. *SDC health policy*. Confederation Swiss.

Galea, S., and R.D. Vaughan. 2019. Health as a means, not an end: A public health of consequence, may 2019. *American Journal of Public Health* 109 (5): 672–673. https://doi.org/10.2105/AJPH.2019.305032.

Godard, B., S. Haddad, R. Huish, and D. Weinstock. 2018. Introduction to ethics and global health. *BMC Medical Ethics* 19 (Suppl 1): 51. https://doi.org/10.1186/s12910-018-0278-1.

Godlee, F. 2011. What is health. *British Medical Journal 343* (d4817): 1. https://doi.org/10.1136/bmj.d4817.
Gore, R., and R. Parker. 2019. Analysing power and politics in health policies and systems. *Global Public Health* 14 (4): 481–488.
Gross, M.L. 2007. From medical neutrality to medical immunity. *AMA Journal of Ethics* 9 (10): 718–721. https://doi.org/10.1001/virtualmentor.2007.9.10.mhst1-0710.
Hall-Clifford, R., and R. Cook-Deegan. 2019a. Ethically managing risks in global health fieldwork. *Health and Human Rights* 21 (1): 7–18.
———. 2019b. Ethically managing risks in global health fieldwork: Human rights ideals confront real world challenges. *Health and Human Rights* 21 (1): 7–18.
Kant, I., and A. Reath. 1997. *Immanuel Kant critique of practical reason*, ed. M. J. Gregor, 1st ed. Cambridge University Press. https://doi.org/10.1017/CBO9780511809576.
Kennedy, J. 2017. How drone strikes and a fake vaccination program have inhibited polio eradication in Pakistan: An analysis of national level data. *International Journal of Health Services* 47 (4): 807–825.
King, H.C., M. Bouvier, N. Todd, C.J. Bryan, G. Montalto, C. Johnson, R. Hawkins, L.A. Braun, J. Malone, and P.W. Kelley. 2019. Shipboard global health engagement missions: Essential lessons for military healthcare personnel. *Military Medicine* 184 (11–12): e758–e764. https://doi.org/10.1093/milmed/usz113.
Koplan, J.P., T.C. Bond, M.H. Merson, K.S. Reddy, M.H. Rodriguez, N.K. Sewankambo, and J.N. Wasserheit. 2009. Towards a common definition of global health. *The Lancet* 373 (9679): 1993–1995. https://doi.org/10.1016/S0140-6736(09)60332-9.
Kusnanto, H., D. Agustian, and D. Hilmanto. 2018. Biopsychosocial model of illnesses in primary care: A hermeneutic literature review. *Journal of Family Medicine and Primary Care* 7 (3): 497–500. https://doi.org/10.4103/jfmpc.jfmpc_145_17.
Lim, M., and D. Blazes. 2015. *Collateral duty diplomacy*. The U.S. Department of Defense and Global Health Diplomacy.
Lincoln, M. 2021. Global health is dead; long live global health! Critiques of the field and its future. *BMJ Global Health* 6 (7): e006648. https://doi.org/10.1136/bmjgh-2021-006648.
Marmot, M. 2018. Inclusion health: Addressing the causes of the causes. *The Lancet* 391 (10117): 186–188. https://doi.org/10.1016/S0140-6736(17)32848-9.
Martinez-Bravo, M., and A. Stegmann. 2022. In vaccines we trust? The effects of the CIA's vaccine Ruse on immunization in Pakistan. *Journal of the European Economic Association* 20 (1): 150–186. https://doi.org/10.1093/jeea/jvab018.
Merritt, M.W., C.S. Sutherland, and F. Tediosi. 2018. Ethical considerations for global health decision-making: Justice-enhanced cost-effectiveness analysis of new Technologies for Trypanosoma brucei gambiense. *Public Health Ethics* 11 (3): 275–292. https://doi.org/10.1093/phe/phy013.
MSF. 2002. *Why is the west ignoring AIDS?* MSF. Médecins Sans Frontières (MSF) International. https://www.msf.org/why-west-ignoring-aids.
———. 2014. *Global bio-disaster response urgently needed in Ebola fight*. MSF. Médecins Sans Frontières (MSF) International. https://www.msf.org/global-bio-disaster-response-urgently-needed-ebola-fight.
———. 2023a. *Where we work*. MSF. Médecins Sans Frontières (MSF) International. https://www.msf.org/where-we-work.
———. 2023b, February. *MSF urges WHO to ensure global health for peace initiative consults all actors*. MSF. Médecins Sans Frontières (MSF) International. https://www.msf.org/msf-urges-who-ensure-global-health-peace-initiative-consults-all-actors.
Pirlea, F., and E. Suzuki. 2023, July 26. *The impact of COVID-19 on global health*. https://blogs.worldbank.org/opendata/impact-covid-19-global-health.
Public Health England. 2014. *Global health strategy: 2014 to 2019*. Public Health England.
PubMed. 2023, July. PubMed. https://pubmed.ncbi.nlm.nih.gov/.
Red Cross. 2011. *Summary of the Geneva conventions of 1949 and their additional protocols*.

Robson, G., N. Gibson, A. Thompson, S. Benatar, and A. Denburg. 2019. Global health ethics: Critical reflections on the contours of an emerging field, 1977–2015. *BMC Medical Ethics* 20 (1): 53. https://doi.org/10.1186/s12910-019-0391-9.

Rubbini, M. 2018. Global Health diplomacy: Between global society and neo-colonialism: The role and meaning of "ethical lens" in performing the six leadership priorities. *Journal of Epidemiology and Global Health* 8 (3–4): 110–114. https://doi.org/10.2991/j.jegh.2017.11.002.

Sauter, M. 2023. Politicized health emergencies and violent resistance against healthcare responders. *Journal of Peace Research* 00223433231158 1. https://doi.org/10.1177/00223433231158144.

Schneider, M.L. 1987. *Health as a bridge for peace*. World Health.

Sibanda, T.S. 2019. *The politicization of healthcare in Europe*. l'Università di Trento.

Smith, R. 2008, July 8. *The end of disease and the beginning of health*. https://blogs.bmj.com/bmj/2008/07/08/richard-smith-the-end-of-disease-and-the-beginning-of-health/.

The Lancet Global Health. 2023. Health-care workers must be trained and retained. *The Lancet Global Health* 11 (5): e629. https://doi.org/10.1016/S2214-109X(23)00172-9.

van Stekelenburg, B.C.A., H.D. Cauwer, D.G. Barten, and L.J. Mortelmans. 2023. Attacks on health care workers in historical pandemics and COVID-19. *Disaster Medicine and Public Health Preparedness* 17. https://doi.org/10.1017/dmp.2022.275.

WHO. 2023c. *WHO global health initiatives*. World Health Organization – Regional Office for the Eastern Mediterranean. http://www.emro.who.int/health-topics/global-health-initiative/index.html.

———. 2023d, May 30. *WHO records more than 1000 attacks on health care in Ukraine over the past 15 months of full-scale war*. https://www.who.int/europe/news/item/30-05-2023-who-records-1-000th-attack-on-health-care-in-ukraine-over-the-past-15-months-of-full-scale-war.

World Bank. 2023. *World Bank open data: Life expectancy at birth, total (years)*. World Bank Open Data. https://data.worldbank.org.

World Health Organization. 2022. Global *Health for peace initiative (GHPI), version 1*. Geneva.

———. 2023. *Global Health and peace initiative (GHPI), version 5*. Geneva. https://cdn.who.int/media/docs/default-source/campaigns-and-initiatives/health-and-peace/v5%2D%2D-ghpi-roadmap.pdf?sfvrsn=29508969_8&download=true.

World Health Organization (WHO). 1948. *Constitution of the World Health Organization*. https://apps.who.int/gb/bd/PDF/bd47/EN/constitution-en.pdf.

———. 2021. *WHO Global Health for Peace Initiative overview*. https://cdn.who.int/media/docs/default-source/campaigns-and-initiatives/health-and-peace/who-global-health-for-peace-initiative-overview.pdf?sfvrsn=4ac4dc31_1&download=true.

———. 2023a. *WHO health workforce support and safeguards list*. https://www.who.int/publications-detail-redirect/9789240069787.

———. 2023b. *Consultations and development of a roadmap*. https://www.who.int/initiatives/who-health-and-peace-initiative/consultations-and-development-of-a-roadmap.

World Medical Association. 2022a. *WMA international code of medical ethics*. https://www.wma.net/policies-post/wma-international-code-of-medical-ethics/.

———. 2022b. *The world medical Association-WMA resolution on humanitarian and medical aid to Ukraine*. https://www.wma.net/policy-tags/war/.

Chapter 14
Risks Associated with Different Understandings of "Medical Neutrality"

Ana Elisa Barbar

Abstract The chapter explores how multiple meanings of "medical neutrality" make it difficult for health care personnel to understand what is expected of them, and how they should behave to respect such neutrality. The article presents the issue using peacetime, emergencies, civil insurgence, and conflict scenarios to discuss what could be the expectations raised by the idea of "neutral behavior" in each case. It assumes two key considerations: one is that, even in conflict settings, neutrality also demands specific behaviors from the health care personnel (beyond the key message of "do not interfere negatively with provision of care", that is clearly a responsibility of parties of conflict). The second is that this chapter does not treat provision of care as a practice that is dissociated of political, social, and economic aspects of the environment where it is being carried out. The risks identified are of individual, professional and systemic nature, and include, *inter alia*, risks of moral distress, violence, loss of protection under International Humanitarian Law. Final recommendations are around the need to provide explicit and contextualized definitions of what is understood as "neutrality" in each context, and to provide training and support to health care professionals, encouraging the debate beyond the humanitarian sector.

1 Introduction

The act of providing health care to another person has been historically associated with sets of values that guide the practice of such caregivers. Despite significant changes in societies, the development of scientific standards or the normative and bureaucratic aspects of fully professionalized approaches to care, the need for attaching responsibility to this set of said "fundamental" ethical principles of care has barely been put to question. What has been up for debate is the meaning of such

A. E. Barbar (✉)
Insecurity Insight, Geneva, Switzerland

D. Messelken, A. E. Barbar (eds.), *Challenging Medical Neutrality*, Military and Humanitarian Health Ethics, https://doi.org/10.1007/978-3-031-69398-4_14

principles in face of contemporary challenges, for example, in debates around conscientious objection or involvement of health care professionals in procedures to terminate pregnancies or bring life to end.

If even the centuries-old ethical principles such as impartiality in care, confidentiality in the professional relation or respect to dignity of the patient are topics of discussion, it is little surprise that a concept borrowed from much more recent debates (the development of "humanitarism" with the Dunantist trend in late nineteenth century) should spark doubts and uncertainty surrounding its meaning. While the recency could justify the debate, it is concerning that a careless use of the expression "medical neutrality" might bring risks to the well-being and integrity of health professionals.

The aim of this article is to explore how the multiple meanings of medical neutrality[1] make it difficult for health care personnel to understand what is expected of them, and how to behave to respect such neutrality. The article will present the issue of "medical neutrality" using peacetime, emergencies, civil insurgence, and conflict scenarios to propose a discussion on what could be the expectations raised by the idea of "neutral behavior" in each case.

In the humanitarian sector, the main frame for action is a set of four core principles, and the respect to such principles is often what distinguishes them from other stakeholders who perform similar activities. The application of said principles as an actionable element in daily work is taken as natural and constantly debated in the sector. Military or "civilian" health personnel have in medical ethics a similar set of values, speaking of the ethos of the profession and to how they perform their duties, but here, the debate around ethical principles is less frequent.

It is much said that the capacity-building offered to most health care personnel regarding medical ethics is quite insufficient given the complexity of situations they might face in their practice (Broussard et al. 2019). Even if it's common that undergraduates in healthcare studies will be presented to a code of ethics[2] and eventually participate in some discussions about the principles and their meaning, there is rarely further *intentional* provocation, in learning or practicing times, to debate the principles and question their application—except for moments when principles are being breached and there's a sanctioning effort to respond to the problem.

The absence of proper capacity-building or routinary discussions does not mean, in any case, that ethical dilemmas are not present in the practice of health care professional. On the contrary, health care professionals are frequently facing ethical dilemmas, that is, situations in which, to decide on a course of action, ethical principles are meant to be carefully weighed and reflected against the background in which the clinical act is taking place. Such delicate situations might happen because of reasons as common as lack of human or material resources in the chain of care,

[1] Many of these meanings and uses were already presented in the preface of this book.

[2] One or more of many, as there's no consensus as to what document is the reference that informs best on the principles to be respected, also considering the multiplicity in health careers.

or might be triggered by specific challenges, such as occurrence of violence or a situation that requires the imposition of strict public health measures.

Inevitably, the practice of healthcare personnel does not remain static and untouched when dilemmas arise, and regardless of what triggered the dilemma in the first place, it is quite likely that it will be by referring to the ethical principles that the issue will be tackled. While for many of the ethical principles existing guidance may support an interested professional in understanding its duties, the lack of a consistency and agreement around the meaning of "medical neutrality" generates very different uses of this term, thus leading to various, sometimes even opposite understandings, and to different expectations of how medical personnel are supposed to act to fulfill this ethical obligation.

As other chapters in this book have presented and debated, the semantic diversity of the concept of "medical neutrality" poses a challenge not only to its interpretation but also to the concrete application. The nuances behind meanings that apply almost exclusively to conflict settings, deriving from International Humanitarian Law (IHL),[3] and imposing to parties of conflict a respectful behavior towards healthcare personnel and objects, is far from being the single interpretation, and other wider-reaching meanings, such as "general respect to human rights" are not hard to come across.

2 Beyond the Clinical Act: Looking into Pressures and Constraints for "Neutral" Behavior

Some uses of "medical neutrality" imply that it corresponds to the righteous application of medical ethics or a broad respect to human rights. Other possible meanings attribute to the term a balanced or a value-neutral approach between opposing forces, that would be achieved by the ethical practice of health care. This balanced approach would be represented by the "non-side-taking" conduct: keeping the primacy of ethical provision of care as the single exclusive activity in which health care personnel engage, leaving behind any element that could be interpreted as bias or favoring to one of the "sides" in question.

This "idealistic" notion of medical neutrality assumes that whatever dispute is at hand, it would be possible to isolate or single-out the provision of health care as dissociated from the dispute. This is firmly based in the belief that the act of care, itself, represents all that there is to healthcare—and excludes the fact that such acts are inscribed into broader social realities, and that what the community understands of a healthcare professionals' behavior may not equal their initial intentions.

[3] Extremely relevant to the understanding of the discussion of the article is that it assumes that "neutrality" has a component of action that is asked of healthcare professionals, not just from the parties to a conflict. The responsibility of parties to respect and protect healthcare personnel, health facilities, medical transport and people in need of care is not open for debate.

The challenge of dealing with "perception" issues is widely debated in the humanitarian sector (Schenkenberg van Mierop 2015).

Another rather non-realistic element of neutrality as value-neutral behavior is that it excludes any degree of personal or individual history, that could eventually affiliate the health care worker to one (or many) disbalancing elements in the equation. Even if the health care personnel behave at best of attempts to preserve a perception of non-side-taking, the reality is that their singular features, their workplace, the eventual employer, the context in which the health intervention is taking place will all be elements considered in how they will be perceived and what neutrality may denote. The brilliant discussion proposed by Mary Turfah in "No Side to Fall in" (2023) speaks of this impossibility of detaching the health provider from the reality where they work.

Because of the complexity of real-life environments, the intention of reducing the issue of neutrality to the clinical act (and often overlapping it with the concept of "impartiality") or to a "middle-point" position in a particular dispute seems to be, at the very least, artificial. Moreover, such simplifications suggest that also the scientific knowledge that produces or encourages health care is static and round, a never-changing world of pre-cooked wisdom that is simply meant to be applied in unison by the health care community. This couldn't be more distant from how health care is effectively produced in daily practice around the globe.

A pragmatic look into the context in which the practice of health care happens points to other considerations that might inform (or divert) a health professional's decision in regards to the "ideal" approach. As individuals, healthcare workers are participating in power dynamics that may be weighing or even determining certain decisions, for example, in highly divided communities where two or more groups do not pacifically coexist. Community-based, peer-based or even hierarchy-based pressure might be a factor that pushes health professionals towards non-neutral decisions, as they are part of that microcosmos and might be ostracized, bullied and even sanctioned if acting as neutral to these pressures (Sharma et al. 2002).

Moreover, personal experiences or values, especially those related to violence or abuse, might bias the approach of health professionals towards non-neutral stances. Situations in which health professionals are confronted with care to perpetrators of violence might also trigger very personal reactions of rejection to ethical principles. While conscientious objection has just recently been incorporated into the International Code of Medical Ethics,[4] the actual incorporation of conscientious objection into health policies that are protective of one's personal biases, yet in alignment with the fundamental principles of healthcare, is far from the reality (Lynöe et al. 2017).

If neutrality is meant to produce a distanced take from other confusing factors in clinical care, healthcare systems are usually far from neutral: they are embedded into political systems, that may have their own stances on if, how and to whom

[4] Very much propelled by contemporary discussions on the proneness of physicians to accept (or not) taking part in procedures to terminate life or pregnancies.

health care is offered (Gostin et al. 2019). While some universal coverage-based systems might be more aligned to the "right to health" as a human right, allowing broad-spectrum (not necessarily free of charge) provision of care to all in need, some health systems defended by governments are not at all "neutral" from a political perspective.

While working under strict or underregulated health systems, health professionals might be faced with policies and pressures such as denying care to undocumented persons, refusing procedures that might facilitate diagnose or reduce suffering in certain health conditions, and even leave diagnosed conditions untreated, which may deteriorate health if uncared for (Physicians for Human Rights and University of Cape Town, 2022, especially section on "Limiting or denying medical treatment or information related to treatment to an individual"). For example, migrants, internally displaced persons and refugees might not have all their documentation read to present, when in need of care. Unemployed persons, or persons falling out of social security systems might also be looped out of healthcare in health systems based mainly in employer-paid insurance to access care. Finally, heavily strained systems such as those in conflict settings and in times of emergencies might become more restrictive and deny certain procedures that, in "normal times" would benefit diagnose and continuous care for chronic diseases (Footer et al. 2018).

Also interfering on a notion of "neutrality" that implies value-neutral decisions is the inevitable tension, in health care, between individual-based "best decisions" and those benefiting the general public. In recent times, the COVID-19 pandemic has shown that decisions against one's individual wish might be frequently the ones that are most protective to the larger population, e.g. when health workers act in actions of mandatory immunization, compulsory isolation, restriction of movement, adoption of restrictive protocols to access care (Downey et al. 2023). While these measures should be fully backed by scientific knowledge to reduce harm to populations, the enforcement of such measures (or the debate around them) comes with arguments and positions that might not be or might not be perceived as neutral in face of human rights or a neutral political perspective.

Given that health care is a human right and that ensuring care for people demands a cascade of bureaucratic and (why not) political choices, the practice of advocating for certain approaches might be perceived, at the individual level, as moving past what neutrality could foresee. As public health advocates, or when defending better and evidence-based practices, medical personnel are often understood to be "non-neutral" as to be pushing for certain agendas and asks that, for some, are inadequate, harmful or exposing gaps and flaws in the health systems (Amnesty International 2020; Hunt et al. 2018).

This last frontier to "neutrality", maybe the biggest challenge in interpreting such a term, is the idea that neutrality is granted by an "exclusively medical" approach to care. Health is, in any case, a political act, whether understood as granting a right to the person in need of care, protecting one from harm, supporting people's dignity and well-being or defending scientific positions as the one that should drive policy and action in care. Making health care an act free from political charge, initiated and finished in the clinical act, is something that seems impossible. Also dissociating the work of individuals from the systems of power and policy in which they operate is not realistic. Some authors go as far as affirming that health personnel, when not humanitarian, cannot be neutral as they are inevitably attached to the government or to another "employer" who'd be, in their turn, governing the choices of the worker (Rubenstein 2014).

Of course, when in conflict, or in situations of violence and other particular disputes, one could argue that the definition of "exclusively medical" is quite straightforward: do not engage in such dispute, and refrain from participating in hostilities. What to do when the dispute at hand is not armed conflict, but around historical racial discrimination, or regarding women's reproductive rights? Can, or should health workers abstain, in order to remain neutral? In the challenge of identifying the proper meaning of "neutrality", medical personnel might find themselves in a tough position when they need to take stances that are often political of nature—and that might yet bring up a perception of politicization of their practice, thus moving beyond the clinical act.

3 What Actions "Neutrality" Calls for?

Although the external scenario seems to be key to understanding what is expected of the healthcare worker in upholding neutrality, the previous section argued that the many variations in interpretation might bring confusion and lead to deep misinterpretation of what neutral behavior could be. The realization of neutrality in peacetime is quite different from what is expected in conflict settings, just as acting "neutral" while an autonomous health professional is different from the "neutrality" that might be possible while working as employee for the national health system. Such nuances in the discussion are often forgotten (Fig. 14.1).

Besides the different implications of the context, the health professional's competences, background and mandate might also bring forward various expectations from the many other groups surrounding the practice of care: patients and their families, the community, security forces, judicial authorities.

The diagram above brings forward a non-exhaustive compilation of possible interpretations that a health professional might adhere to, when enacting the principle of neutrality in its daily routine. The arbitrary divisions are meant to display blunt differences depending on context and affiliation, and while there could be many other possible distinctions, these categories are sufficient to sustain the debate on risks that may rise from such diverse interpretation.

Note that, in many scenarios, interpretations of neutral behavior can be conflicting or antagonizing if co-existing in the same context, giving room to potential misperceptions of the alignment of healthcare personnel to best ethical practices. In peacetime and some types of emergencies, the main lead to enacting neutrality seems to be a) respect to human rights and general healthcare ethics principles, and b) avoid harm in policy and clinical acts. On the opposite end of the spectrum, the

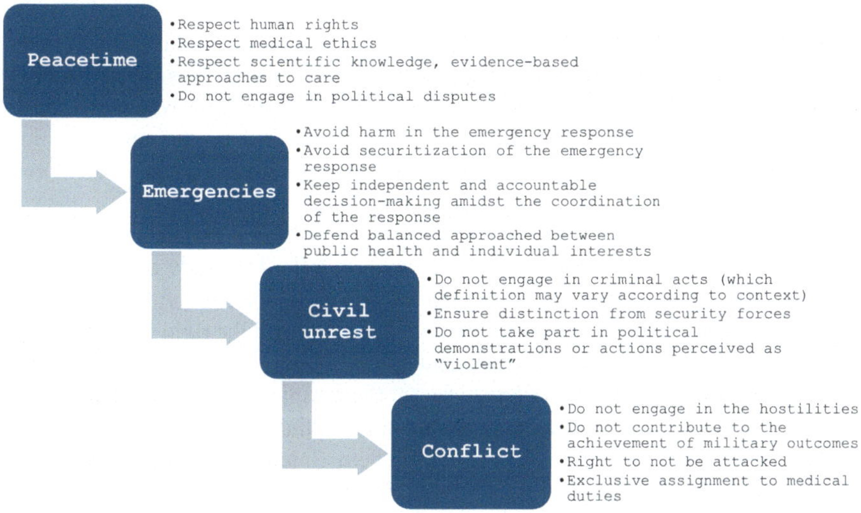

Fig. 14.1 Possible meanings of "medical neutrality" in different scenarios, as identified in the various documents reviewed in this chapter

driving force of contexts where violence and social unrest is at play (conflict, internal disturbances, social uprising) seems to be the key message of IHL's neutrality[5]: "do not engage in hostilities", which in some cases might be shaped into "do not engage in criminal acts" or even "maintain distinction from security forces".

4 Risks Associated to Multiple Interpretations and Misperception

The lack of clarity around the principle of neutrality and its denoting behavior is not free from consequences. While one could argue that it is simply a matter of respecting medical ethics and avoiding securitization of care or engagement in hostilities, the reality is that, for a healthcare professional, certain aspects of neutrality can be quite challenging to understand and uphold.

At a personal level, the confusion of what is expected of a professional when upholding ethical principles may lead to moral distress, and it is widely demonstrated that this emotional burden is especially high in situations of emergencies and other extraordinary conditions such as working in conflict and civil unrest settings. Also, lack of common definitions might give room to free interpretation that can land far from what could be understood as acceptable in that context, by the

[5] For a detailed explanation of neutrality according to International Humanitarian Law, please see other chapters of this book.

community. Finally, a sense of loss of alignment to ethical principles might be felt if a professional feels to be disrespecting neutrality, even if for the sake of upholding other core values of medical ethics: for instance, when deciding to advocate in favor of inclusiveness (Keshet and Popper-Giveon 2017) or when preventing security forces from carrying out potentially harmful act against patients, healthcare professionals might personally question whether they are not obliterating the issue of neutrality in care (Hathout 2012).

Another important risk of neutrality's varying definition is that of acting in ways that might imply in loss of protection according to IHL. This body of law foresees that the clinical act must be impartial, and the behavior of the practitioner must not favor any military party. The idea of medical neutrality and the right to be protected stems in great deal from this exclusive dedication to medical functions. But in interpreting what such medical functions mean, or the meaning of "favoring a party of a conflict" as supporting a particular military gain, healthcare professionals might not be fully complying to what is expected of them (intentionally, or when unconsciously fulfilling hierarchy- or policy-based instructions).

The broad and largely accepted indication is that engaging in hostilities in any way is the most obvious non-neutral behavior. Yet, even in conflict settings, the simple explanation of neutrality as being an expectation to "not engage in hostilities" seems to be too superficial, as several nuances might be drawn from such contexts (Alhoff and Potts 2018; Turfah 2023), shifting the perception from external observers towards a personal interpretation of "adherence to" or "lack of" neutrality: is the medical staff engaged and paid by the government who is a party of the conflict? Are the health care practices concentrated in specific areas which are occupied or dominated by one of the parties? What is the realistic possibility of healthcare professionals to access and treat wounded persons without facing persecution? (Rubenstein and Haar 2022) Is there a call for defending the provision of health care with anything beyond light weapons, in line with recent discussions around "extended self-defence" and use of embedded medical teams?

From the medical practice itself, a partial approach to care (selective entry doors to care, preferences based on military allegiances or unequal clinical protocols) could well represent forms of "favoring" a party in a context of conflict. Also, many worrisome approaches to "counterterrorism", suggest that material support of any kind, including provision of healthcare to certain groups is a way of backing these individuals, condemning actions that would be, in theory, aligned to medical ethics (Médecins Sans Frontiers 2021).

On the one hand, the factual loss of protection is a legal interpretation that has specific implications, including the need to officially and timely notify the opposite party of such reading, before any offensive action. On the other hand, the perception of lack of neutrality, even if the health professionals are behaving at best practices, might endanger their safety and integrity regardless of a "legalistic" or "lawful" interpretation.

Threats, persecution and attacks are not perpetrated only if medical personnel are acting against neutrality. In fact, such terrible events might happen based on misperception that healthcare professionals are prioritizing a specific group, or because they are perceived to be acting in favor of certain non-medical outcomes (political, religious or military). This reality shows that, regardless of intentionality, the way health workers are perceived to be acting as neutral (or not) may enhance risks of violence against them (Wise et al. 2021). This sheds light on the importance of perception, beyond the immediate act. It also demonstrates the stark need to understand the principle of neutrality in ways that others, too, can commonly agree and respect.

Differently from humanitarian health workers, who are normally in a constant attempt to prioritize balanced approaches in humanitarian response, local health workers might be inevitably caught in a very unbalanced scenario simply because they happen to live or work in that area prior to the conflict (Broussard et al. 2019; Turfah 2023). In such conditions, these health professionals might be offering care in larger scale or almost exclusively to one side of society, and even when this has no links with choices made directly by the health personnel, they might be perceived as non-neutral given the context (Orr and Fleming 2023). The same thing happens with eventual negative perception in the association of health care facilities and personnel to security forces or to the governing authorities, when these people are seen exchanging information or sharing spaces.

The "securitization" of healthcare, or the mere presence of security forces and agents in health care facilities might be extremely detrimental to perception of neutrality. If neutrality is to be interpreted as "not engaging in hostilities", this should not be a problem, yet in different cases health workers were accused of lack of neutrality by the fact of allowing the presence of such security actors in healthcare sites, or because health teams used the support of security agents to gain access or protect them in certain insecure environments (Sauter 2023). This idea of association as indicative of non-neutrality is especially challenging for military medical personnel, who are, by principle, part of an armed force. Moreover, military medical personnel are often accompanying military engagement to provide medical support to the engaging troops, therefore they will hardly be seen as "uninterested party" (Physicians for Human Rights and University of Cape Town 2002, especially section on "Denial of care for political reasons and during armed conflicts"). For these health professionals and military medical corporations, adherence to medical ethics (both at clinical and policy levels) and refraining from any engagement in hostile actions seem to be indeed the single avenue of accountability on the issue of neutrality.

For all mentioned groups—civilian, humanitarian or military health professionals—the consequences of loss of trust from the community and/or lack of access to the people in need of care is another clear risk associated to misperceptions of what neutral behavior means. When community leaders and health responders don't have the same understanding on upholding neutrality, the former might act as gatekeepers and prevent access of the latter to the population (Dawkins et al. 2021; Thompson 2015). They might also influence health-seeking behavior of community members. In a worse case scenario, the disagreement on the meaning of neutrality might result in official refusals by governmental authorities for humanitarians and other health actors to move or work in certain areas, as a response to what they might understand as biased or criminalized by local legislation.

As the loss of trust and reduced access will consequently impair the health response itself, it is possible to argue that the varying definitions of medical neutrality might unintendedly weaken the efficiency in achieving the intended health outcomes. Moreover, in times of social media and mis- or disinformation digital battles, all public communication and public reaction to positions interpreted as non-neutral might be very detrimental to individuals and organizations (Insecurity Insight 2023), from the adherence and trust that can be built with target groups of the health response, to the safety and security of the health teams while delivering care.

Finally, the adoption of policies and clinical approaches strongly associated with political agendas, (e.g. sexual and reproductive health, delivery of preventive therapeutics such as vaccines, delivery of care for refugees, migrants or other vulnerable population) with potential perception of non-neutrality due to heavy politicization of the debate, might bring two detrimental effects to the health response: firstly, as mentioned to other concerns, there might be reduced possibility to deliver care or access the population in need of it. Second and most importantly, the reactions against such alleged lack of neutrality might surface proposals of healthcare delivery that infringe human rights, cause harm or that are negligent regarding scientific and evidence-based knowledge (Hewlett and Hewlett 2008, particularly in Chapter 4, section "Origin stories, history and politics").

To summarize the risks described in this section, personal risks to the health professional range from moral distress to a sense of ethical overriding that can be detrimental to one's practice and well-being. At the level of the health response, some risks are related to safety and security and to the protection granted to the medical mission in conflict settings, while other risks relate to loss of trust and reduced access to people in need of care. Finally, in a broader social perspective, risks of misunderstanding the concept of neutrality might result in chronic problems of negative perception towards health workers and reduce adherence to approaches backed by scientific knowledge and human rights.

5 Conclusion

For this chapter, there are two important considerations regarding the understanding of neutrality: one is that, even in conflict settings, neutrality also demands specific behaviors from the health care personnel (beyond the key message of "do not interfere negatively with provision of care" that is clearly a responsibility of parties of conflict). The second is that this chapter does not treat provision of care as a practice that is dissociated of political, social, and economic aspects of the environment where it is being carried out.

Agreeing with the principles of medical ethics, especially in that the medical function should not be used to cause harm, and should be accessible, fairly distributed to all in need of care, this chapter proposes an emphasis on the ownership and accountability of health care personal regarding their professional choices, with a strongly actionable approach to "medical neutrality". Of course, part of the discussion will be around limitations in this capacity for action, and risks thereof. This approach does not exclude the responsibility of others (weapon-bearers, political or technical decision-makers, community members, or any other stakeholder) to be supportive of conducive environments for the provision of care, keeping health care free from violence, supporting constructive debates and action, and avoiding harm at all levels of care.

As argued in this article, the various meanings associated with "neutrality", as well as the lack of clarity regarding what is the expected behavior of healthcare personnel to uphold it, are not free of risks. It seems to be insufficient the guidance that simply recommends "neutral behavior" or "respect neutrality", as such concepts might mean different things to different people, potentially hindering the outcomes of the health response and at times even putting the integrity of the health care personnel in danger. The reckless use of the term "neutrality" does not support the daily adoption of what the ethical principle seems to preach: a clinical approach that is evidence-based and that respect medical ethics and human rights, with strong rejection of participation, by health care professionals, in any act of violence or harmful act.

While it is far beyond the scope of this article to propose an operational definition to medical neutrality that could eventually be applied in the so many discussed scenarios, the main recommendations stemming from the presentating risks associated with the semantic diversity are: clearly define what neutrality means for a particular health organization; propose and support debates with communities, authorities, religious and academic representatives on what are the nuances behind application of neutrality in given contexts; provide support and guidance to health practitioners in what are best reflections and best practices when faced with ethical dilemmas.

An extended discussion of neutrality meanings and action, beyond the realm of strict humanitarian action, is needed, including the practice of military and civilian healthcare professionals. Other recommendations go in the direction of avoiding the use of "neutrality" to denote other ethical principles such as "neutrality" or "respect to dignity of the patient", as well as avoiding the paradox of calling for value-neutral approaches while encouragement to whistleblowing against harmful practices is (thankfully) on the rise.

Finally, it is very much needed that health care actors (again, beyond the humanitarian practice) become fluent in the factors that surround securitization of healthcare, the complex relationship between law enforcement and delivery of healthcare, public health and human rights. This may enable a deeper and nuanced attitude to operationalize neutrality in contexts beyond peacetime, where the life, well-being and work of health professionals may be linked to a positive perception of their "neutral" behavior.

References

Alhoff, F., and K. Potts. 2018. Medical immunity, international law and just war theory. *Journal of the Royal Army Medical Corps* 0: 1–10. https://doi.org/10.1136/jramc-2018-001020.

Broussard, G., L. Rubenstein, C. Robinson, W. Maziak, S.Z. Gilbert, and M. DeCamp. 2019. Challenges to ethical obligations and humanitarian principles in conflict settings: A systematic review. *Journal of International Humanitarian Action* 4: 15. https://doi.org/10.1186/s41018-019-0063-x.

Dawkins, B., C. Renwick, T. Ensor, B. Shinkins, D. Jayne, and D. Meads. 2021. What factors affect patients' ability to access healthcare? An overview of systematic reviews. *Tropical Medicine and International Health* 26 (10): 1177–1188. https://doi.org/10.1111/tmi.13651.

Downey, E., H.S. Fokeladeh, and H. Catton. 2023. *What the COVID-19 pandemic has exposed: The findings of five global health workforce professions*. Geneva: World Health Organization.

Footer, K.H.A., E. Clouse, D. Rayes, Z. Sahloul, and L. Rubenstein. 2018. Qualitative accounts from Syrian health professionals regarding violations of the right to health, including the use of chemical weapons, in opposition-held Syria. *BMJ Open* 8: e012096. https://doi.org/10.1136/bmjopen-2017-021096.

Gostin, L.O., J.T. Monahan, J. Kaldor, M. DeBartolo, E.A. Friedman, K. Gottschalk, et al. 2019. The legal determinants of health: Harnessing the power of law for global health and sustainable development. *The Lancet Commissions* 393: 1857–1910. https://doi.org/10.1016/S0140-6736(19)30233-8.

Hathout, L. 2012. The right to practice medicine without repercussions: Ethical issues in times of political strife. *Philosophy, Ethics and Humanities in Medicine* 7: 11. https://doi.org/10.1186/1747-5341-7-11.

Hewlett, B.S., and B.L. Hewlett. 2008. *Ebola culture and politics: The anthropology of an emerging disease*. Belmont: Thomson Wadsworth.

Hunt, M., A. Chénier, K. Bezanson, E. Nouvet, C. Bernard, et al. 2018. Moral expériences of humanitarian health professionals caring for patients who are dying or likely to die in a humanitarian crisis. *Journal of International Humanitarian Action* 3: 12. https://doi.org/10.1186/s41018-018-0040-9.

Insecurity Insight. 2023, July. *Sudan's doctors falsely accused of siding with the rapid support forces*. Social Media Monitoring and report, Geneva.

Keshet, Y., Popper-Giveon, A. February 2017. Neutrality in medicine and health professionals from ethnic minority groups: The case of Arab health professionals in Israel. Social Science and Medicine, 174:35–42. https://doi.org/10.1016/j.socscimed.2016.12.019.

Lynöe, N., J. Björk, and N. Juth. 2017. Is healthcare providers' value-neutrality depending on how controversial a medical intervention is? Analysis of 10 more or less controversial interventions. *Clinical Ethics* 12 (3): 117–123. https://doi.org/10.1177/147775091770415.

Médecins Sans Frontières. 2021, October. *Adding salt to the wound*. MSF Counterterrorism Report, Brussels.

Orr, Z., Fleming, M. D. February 2023. Medical neutrality and structural competency in conflict zones: Israeli healthcare professionals' reaction to political violence. Global Public Health, 18(1), doi:https://doi.org/10.1080/17441692.2023.2171087.

Physicians for Human Rights and University of Cape Town (School of Public Health and Primary Health Care). 2002. *Dual loyalty & human rights in health professional practice—Proposed guidelines and institutional mechanisms*. Report, Boston.

Rubenstein, L. 2014. A way forward in protecting health services in conflict: Moving beyond the humanitarian paradigm. *International Review of the Red Cross* 95 (890): 331–340. https://doi.org/10.1017/S1816383113000684.

Rubenstein, L., and R. Haar. 2022. What does ethics demand of health care practice in conflict zones? *AMA Journal of Ethics* 24 (6): E535–E541. https://doi.org/10.1001/amajethics.2022.535.

Sauter, M. 2023. Politicized health emergencies and violent resistance against healthcare responders. *Journal of Peace Research* 0 (0). https://doi.org/10.1177/00223433231158144.

Schenkenberg van Mierop, E. 2015. Coming clean on neutrality and independence: The need to assess the application of humanitarian principles. *International Review of the Red Cross* 97 (897/898): 195–318. https://doi.org/10.1017/S181638311500065X.

Sharma, G.K., B. Osti, and B. Sharma. 2002. Physicians persecuted for ethical practice in Nepal. *The Lancet* 359 (9316): 1519. https://doi.org/10.1016/S0140-6736(02)08422-2.

Thompson, A. 2015. Humanitarian principles put to the test: Challenges to humanitarian action during decolonization. *International Review of the Red Cross* 97 (897/898): 45–76. https://doi.org/10.1017/S1816383115000636.

Turfah, M. 2023. No side to fall in: Medical neutrality in Gaza. *The Baffler*, December 15. Accessed lastly June 13 2024, in: https://thebaffler.com/latest/no-side-to-fall-in-turfah.

Wise, P.H., A. Shiel, N. Southard, et al. 2021. The political and security dimensions of the humanitarian health response to violent conflict. *The Lancet* 397 (10273): 511–521. https://doi.org/10.1016/S0140-6736(21)00130-6.

Chapter 15
Medical Neutrality and Political Engagement

Leonard Rubenstein

Abstract A claim is often made that medical neutrality requires that health workers practicing in conflict settings or in situations of political violence refrain from taking a position on controversies of a political, racial, religious, or ideological nature. The admonition is also applied to human rights documentation and reporting and advocating for political change. That position, however, is highly problematic, in part because the phase is "medical neutrality" is not contained in international law and because has at least three different meanings: immunity from attack, impartiality, and political neutrality. More importantly, refraining from political engagement is inconsistent with contemporary medical ethics, including obligations to prevent abuses, report those that take place, and advance health equity. Moreover, the humanitarian principle of humanity, on which the purported obligation is predicated, is not morally grounded and lacks the status of the companion principles of humanity and impartiality. These principles both justify and encourage health professionals to engage in political matters that can protect the rights and health of patients and the larger populations. In taking such stances, health professionals can employ decision-making processes that account for possibly conflicting values and obligations and assessing the benefits, burdens, and risks of alternative courses of action.

In January 2020, after a military coup against Myanmar's elected government, doctors joined in the leadership of a movement that sought to restore democracy. They marched in the streets in white coats and boycotted public hospitals. In a letter to *The Lancet*, A group of physicians explained their actions, especially the hospital boycott, in light of their ethical obligations to patients and the larger community:

L. Rubenstein (✉)
Johns Hopkins Bloomberg School of Public Health, Center for Public Health and Human Rights, Baltimore, MD, USA
e-mail: lrubenstein@jhu.edu

D. Messelken, A. E. Barbar (eds.), *Challenging Medical Neutrality*, Military and Humanitarian Health Ethics, https://doi.org/10.1007/978-3-031-69398-4_15

> Our duty as doctors is to prioritise care for our patients – but how can we do this under an unlawful, undemocratic, and oppressive military system? For emergency care providers, limiting access to life-saving interventions presents an acute and complex ethical challenge, notwithstanding the significant risks to the public. 50 years of previous military rule failed to develop our health system and instead enshrined poverty, inequality, and inadequate medical care. We cannot return to this situation (Soe et al. 2021).

The writers added that doctors provided emergency services in private and charity hospitals and staffed ambulances and clinics in the street but recognized that these efforts could not ameliorate the impacts of the boycott on patient care. Even so, they believed that they had an ethical duty to avoid potentially grievous harm to the people of Myanmar and the collapse of health services, which in their judgment outweighed any harms from the boycott.

They were hardly the only health workers in recent years to protest, and often lead, action against repression, state violence, and other human rights violations. In 2019, Sudanese doctors played a major role in a pro-democracy movement, as did health workers in Arab Spring protests in 2011 in Bahrain, Egypt, and Syria. For more than two decades, Turkish doctors campaigned against torture of detainees and risked prosecutions for reporting evidence of torture in medical exams they conducted. (Rubenstein 2021, ch. 3) In many other armed conflicts and situations of political violence, health workers have reported human rights violations and demanded accountability for them, and not only in cases where they were, as now all too common, the targets.

These activities were based on their beliefs that they had an ethical as well as a political obligation to protect the human rights of people they served. In taking this stance, these health workers departed from the neutral stance reflected in admonitions to humanitarian groups adopted by the UN and aid organizations that they not engage "in controversies of a political, racial, religious or ideological nature." (UN OCHA 2012). Although these instructions are silent on whether they apply to the vast majority of health workers not affiliated with a humanitarian aid group—such as those protesting in Myanmar, Sudan, and Syria—it is often asserted that they have an ethical obligation of "medical neutrality" (ICRC 2014; Mehring 2014). This chapter addresses that claim, the principles and values that underlie it, and the process for resolving potential conflicts among obligations to patients, organizations, and populations. Put more simply, were the actions of the doctors in Myanmar and elsewhere a violation or a professional obligation or a fulfillment of one?

To answer the question, it is necessary to review the multiple meanings of medical neutrality and the ethical confusions attendant to those meanings. It is also essential to understand whether one of the purported duties of medical neutrality—to refrain from political engagement or expressions of ideological nature, and to limit reporting on human rights violations in conflicts—squares with the ethical obligations of health workers to advance the health and rights of patients and populations, as those duties have evolved in recent decades.

1 Original Meanings: Neutrality and Health Care in the Nineteenth Century

More than 30 years ago, Hans Kalshoven, an astute and influential scholar on the laws of armed conflict, wrote "one will look in vain in all the conventions on international humanitarian law in force" for the term "medical neutrality." He added: "Nor is the term free from ambiguity" (Kalshoven 1992). He was right. There are at least three different meanings of the phrase, each one derived from a different source and tradition: The immunity of health workers in armed conflict and other situations of violence from attacks; the obligation of impartial care for the wounded and sick; and the claimed duty to refrain or severely limit political engagement and speech and human rights reporting.

1.1 Neutrality as Immunity

The first, and most commonly understood meaning of medical neutrality is that, in caring for the wounded and sick in war, health workers are immune from attack or interference. The idea derives from the original Geneva Convention of 1864. It deemed health workers and civilians "who bring help to the wounded," along with military hospitals, homes where civilians cared for the wounded, and medical transports, "shall have the benefit of neutrality," which required protection and respect (Geneva Convention 1864).

The World Medical Association and other medical and nursing groups, NGOs, and human rights organizations, regularly employ the phrase "medical neutrality" to convey this immunity. In 2011, for example, the WMA issued a statement that "Physicians and other health care personnel must be considered as neutral in armed conflicts and situations of violent civil unrest. They must not be prevented from fulfilling their ethical, moral, and professional duties" (World Medical Association 2011).

The use of the word "neutrality" in the 1864 Convention, however, is odd, even in the context of its time, as neutrality in war referred to the duties of non-combatant states, not individuals. In the mid-eighteenth century, the renowned and highly influential international lawyer Emmerich de Vattel described neutrality in war in a 1758 treatise on the law of nations. He identified two obligations of states claiming to be neutral. First, they must refrain from participation in the conflict, whether through direct military action or material, financial, or territorial support for belligerents. Second, they could engage in commerce with belligerents, but must act impartially with respect to it. If they conducted their affairs in these ways, neutral states were not subject to invasion or attack by a belligerent (Kenny 2015). By the end of the nineteenth century, that meaning of neutrality was incorporated into international conventions (Kalshoven 1992).

That concept of neutrality was thus a poor fit for medical activities during nineteenth century wars. Although Henri Dunant, whose writings inspired the Convention, promoted the creation of voluntary societies to attend to the wounded and sick on the battlefield, at the time most health workers offering care in war were members of a nation's military. As part of a country's war machinery, their roles included preserving and strengthening the capacity of the nation's fighting forces. Indeed, the 1864 Geneva Convention required military hospitals in all circumstances to display their national flag along with the newly established emblem, a red cross or, later, a red crescent.

The concept of neutrality was also superfluous as a foundation for the obligation in the Convention to protect and respect health workers. Those requirements could stand on their own under the principle of humanity, to relieve the suffering of human beings, whoever they are or whatever they did, that was central to Dunant's vision and remains the core value of the Geneva Conventions to this day. Dunant's memoir of the battle that inspired his work, *A Memory of Solferino* (1959), was animated by his anguish at the suffering of wounded soldiers on both sides, and eloquently conveyed unspeakable, gruesome injuries of the wounded soldiers, and their desperate need of water and solace, to the point that they sometimes begged for death. He glorified the dedication of doctors and local citizens whose devotion to humanity led them to work tirelessly to try to relieve that suffering and making "no difference between friends and enemies" (Dunant 1959). He saw them as "noble and compassionate spirits" acting "in a spirit of peace, for the purpose of comfort, from a spirit of self-sacrifice" (Dunant 1959).

He especially celebrated the French military doctors who, he wrote, "not only did everything that was humanly possible without distinction of nationality; they grumbled and complained at their inability to do more" (Dunant 1959). For Dunant, then, the health workers did not conduct their medical activities based to whatever political or military affiliation they had. Rather, their work was guided only by the need to serve humanity. It was because of that commitment, as he perceived it, that he put them on a pedestal with the most courageous soldiers: "Surely if those who make the slaughter can claim a place on the role of honour, those who cure often at risk of their lives, are entitled to their due of esteem and gratitude" (Dunant 1959).

1.2 Medical Neutrality as Impartiality

The second meaning of medical neutrality is impartiality, a duty of parties to the conflict to provide care based on need and prohibiting discrimination or differentiating care based a person's ethic or national identity, religion, tribe, gender, military affiliation, or political views. The 1864 Geneva Convention required that wounded and sick combatant, "of whatever nations they may belong" be collected and cared for (Geneva Convention 1864)." In other words, parties to the conflict are required to provide care based on medical criteria alone, to friend and foe alike. That concept has been articulated through every iteration of the Conventions, including their expansions from international armed conflict to civil wars.

The duty of impartiality set out in the Convention only applied to individual health providers insofar as they are agents of a belligerent in the conflict. It was only much later that a duty of impartial care became embedded in the ethics of the health professions as well as in principles governing the work of humanitarian organizations.

1.3 Medical Neutrality as Refraining from Supporting One Side in a Conflict

The third meaning of medical neutrality is not taking a political or ideological stance, which includes naming perpetrators of human rights violations during a conflict. The question did not arise at the time of the drafting of the Geneva Convention of 1864. As noted, the legal meaning of neutrality was limited to a neutral state's duties to refrain from militarily support of a belligerent in an armed conflict, to prohibit use of its territory from use by one of the parties to the conflict, and not to supply weapons to belligerents. A neutral state, moreover, was not obligated to take action to prevent its nationals from supplying weapons, nor to outlaw the use of private communication channels and equipment that would aid a belligerent's cause. (DLP Forum 2023). Nor were political statements in favor of one party or against another barred. These limitations on the meaning neutrality remain in place case today. For example, the U.S. Department of Defense Law of War Manual states that "merely expressing sympathy for a belligerent's cause would not cause a person not otherwise taking part on the war to forfeit one's neutral character" (U.S. Department of Defense, rev. 2023, Para. 15.6.2.2.).

2 Twentieth Century Corrections and Further Confusions

During the twentieth century, multiple revisions to the Geneva Conventions expanded protection of health care from attack and clarified the basis for medical immunity. Crucially, neutrality was eliminated as the grounding for that immunity. By the 1960s however, the idea of political activities as a breach of humanitarian and medical neutrality took hold.

2.1 Elimination of Neutrality as Basis for Immunity of Health Care from Attack

The confusion of invoking neutrality as a basis for immunity of health providers in armed conflict took decades to correct. In 1906, however, Louis Renault, an international law scholar and diplomat, who later won the Nobel Peace Prize, argued that it was unrealistic for military-affiliated or citizen medics adjunct to an army to be considered neutral. Rather, he said, they were part of or allied to an organization pursing military objectives. As such, Renault pointed out, they were enemies, but "enemies with a special task, and protection and special immunities must be accorded to them precisely to enable them to perform that task." (Kalshoven 1992). His arguments persuaded delegates revising the Geneva Convention that year to remove reference to neutrality as a rationale for the duty imposed on combatants to respect and protect health care. Instead, the obligations to "protect" and "respect" the wounded and sick and their caregivers stood of their own, grounded in the principle of humanity. In the context of medical care in armed conflict, neutrality never reappeared in subsequent iterations of the Conventions, while duties of combatants to protect and respect the wounded and sick, medical personnel, and health facilities remained.

2.2 The Protection of and Breaches of Medical Impartiality

Over the course of the twentieth century, the concept of impartiality became embedded in the values of humanitarian organizations and medical ethics, gradually overcoming long entrenched racism and colonial attitudes in Western medicine. As the Geneva Conventions evolved, moreover, provisions were added to protect individuals engaged in medical activities from punishment for carrying out their work in accordance with medical ethics. The process began with the 1949 Conventions. During and after World War II, health professionals were punished, even prosecuted for treason, for treating wounded enemies, especially resistance fighters. In its aftermath, the First Geneva Convention of 1949 added a provision that "no one may be molested or convicted for having nursed the wounded or sick." (Geneva Convention 1949, art. 18). The two 1977 Protocols to the Geneva Conventions expanded protection by prohibiting punishment of any person engaged in health care for adhering to their ethical duties or requiring them to act in contraventions of these duties. (Protocols Additional to Geneva Conventions 1977). Human rights law similarly protects impartiality (UN General Assembly 1991).

In armed conflict, moreover, a civilian health professional's breach of the duty of impartiality does not, of itself, abrogate their immunity from attack. Breaches of medical ethics such as violation of the duty of impartiality are typically addressed through professional disciplinary procedures. Under the law, protection of civilian health workers is lost only if they directly take part in hostilities, a standard even more strict than the rule—"committing acts harmful to the enemy" —that applies to individuals specifically designated as medical personnel by the parties to the conflict. (Kolb and Nakashima 2019). These distinctions are, however, often lost on combatants, as health workers are attacked or arrested for providing impartial care to enemies (Rubenstein 2021).

2.3 *Emergence of Political Neutrality as an Expectation for Health Care in Conflict*

Health workers in war zones and situations of political violence are commonly not politically neutral. As Renault pointed out, military physicians and nurses have sworn allegiance to their country. Similarly, health workers who are members of an ethnic, religious, political, or national or group that is under assault cannot shed their identity or allegiance, nor be expected to do so. Even if they asserted their neutrality, their enemies would not likely accept their claim. These health workers, however, generally embrace the obligation of impartiality even when under assault (Rubenstein 2021).

As noted above, the legal and diplomatic meaning of neutrality is restricted to actions and policies that directly aid the military objectives of a belligerent, not to political beliefs or activities. In the past half-century, however, as expectations of humanitarian neutrality in war emerged, it was interpreted to require aid groups, including those offering medical care, to refrain from taking political positions or speaking out on human rights violations in armed conflict. By extension, this expanded conception of neutrality has been applied, quite problematically, to civilian health workers unaffiliated with humanitarian organizations. (Mehring 2014; ICRC 2014).

This meaning of humanitarian political neutrality—and its incorporation in the meaning of medical neutrality — is of recent vintage. At least through World War II, the ICRC's assertion of its neutrality was more of a label rather than a core principle. Neutrality was not among four principles articulated in 1875 by Gustave Moynier, Dunant's colleague and co-founder of the International Committee for Relief to the Wounded (later re-named International Committee of the Red Cross (ICRC)) (ICRC 1996). Nor was neutrality included in the 1921 Fundamental Principles of the Red Cross. (ICRC and IFRC 2020). During the pre-war period, moreover, national Red Cross societies identified strongly with their countries' nationalist and even military aims, most notoriously in Fascist Italy and Nazi Germany (Steinacher 2017). In 1942, the Swiss military, with approval of the Swiss government, sent Red Cross doctors, technically under the auspices of the Swiss Red Cross, to the eastern front to aid the campaign against the Soviet Union. The doctors were bound by German military law and were forbidden to treat wounded Soviet soldiers or Russian civilians (Steinacher 2017).

That decision reflected the ICRC's conception (and that of the Swiss Red Cross) of deep ties to the Swiss government. The ICRC was staffed by Swiss nationals, many of whom had close ties with Swiss officials and often rotated between government service and ICRC leadership. The ICRC also identified with and followed Swiss foreign policy, which historian Gerald Steinacher has shown contributed significantly to its moral compromises during the war, especially its response to the Holocaust. (Steinacher 2017). Moreover, during and after the war, ICRC leaders participated in activities that were overtly political. For example, Jacob Burkhardt, its Vice-President at the time, engaged in secret peace talks between the Third Reich and England. At the Nuremberg trials, moreover, he voluntarily testified on behalf of two Nazi defendants. One of them, Ernst Kaltenbrunner, was an SS leader who played a key role in planning and carrying out the murder of Jews. (Notwithstanding Burkhardt's testimony, Kaltenbrunner was found guilty of crimes against humanity and executed) (Steinacher 2017).

The conduct of the ICRC and Red Cross societies during the war, however, led to both defensiveness and soul-searching—but even then, not to embrace of political neutrality as a fundamental principle. In 1946, the board of governors of the League of Red Cross Societies, the predecessor to the International Federation of Red Cross and Red Crescent Societies, adopted a new set of 13 governing principles to add to its pre-existing principles of impartiality; political, religious, and economic independence; universality of the Red Cross; and equality of the National Red Cross Societies. Neutrality was not among these revised principles (League of Red Cross Societies 1946). Two years later, Max Huber, who had been president of the ICRC from 1928 to 1944, published an article articulating the ICRC's principles. He briefly referenced neutrality, but only to emphasize the significance for ICRC's work of its connection — geographically and by its governance — to Switzerland, which of course was neutral in the conventional legal sense. Huber also noted that the ICRC was neutral, but the context suggests that he equated neutrality with impartiality (Huber 1948).

Eventually, though, Jean Pictet, the renowned ICRC lawyer, who played a seminal role in the drafting of the 1949 Geneva Conventions and, later, the 1977 Additional Protocols, wrote a treatise on the ICRC's principles. Published in 1955, Pictet's volume led the ICRC to appoint a commission to review them. After a long process, in 1965, ICRC adopted new fundamental principles including, for first time, political neutrality. The new document stated that "in order to continue to enjoy the confidence of all, the [Red Cross and Red Crescent] Movement may not take sides in hostilities or engage at any time in controversies of a political, racial, religious or ideological nature" (ICRC and IFRC 2020).

Pictet followed the adoption of the principles with a commentary on them. He brought passion, eloquence, and sometimes overheated rhetoric in defense of the now-expanded meaning of neutrality that precluded taking a political stance and remaining silent on controversial issues. He recognized that neutrality was not grounded in moral principles, as humanity and impartiality are, but rather is a pragmatic means to fulfill those obligations: "to reach people who need help in crises and to maintain a dialogue with those involved in armed conflict and other violence." He wrote that neutrality "opens doors to access, services, and dialogue with all parties." (Pictet 1979). The understanding that neutrality is pragmatically, not morally, based, remains today (Labbe and Daudin 2016).

Pictet was especially passionate in calling for political neutrality, which went far beyond the idea of neutrality as previously understood. He warned of terrible consequences if the Red Cross movement engaged in political expression or called out human rights violations. Doing so, he said, would "compromise its universal character." Maintaining neutrality, on the other hand, would be "a sign of its serenity, of fidelity to its ideal." He added that "Any ideology to which the Red Cross might attach itself could only serve to diminish its freedom of action and reduce its objectivity." The neutral man, he wrote "refuses to make a judgment." Pictet's hostility to political engagement reached the visceral. The commentaries argue that "The world of politics is characterized by struggles which reach the pitch of savagery; it is not only rival interests which are in conflict, but even sincere advocates of social progress of this, that, or the other party." As a result, he wrote, "Red Cross institutions must beware of politics as they would of poison, for it threatens their very lives. Politization undoubtedly constitutes the greatest danger now confronting the Red Cross" (Pictet 1979).

Such political involvement contrasts, he said, contrasts with "humanitarian activities which help to promote mutual understanding and goodwill among nations whatever their political differences." (Pictet 1979). Pictet acknowledged that neutrality was not always easy, but wrote that it was essential and can even lead to peace:

> It demands real self-control; it is indeed a form of discipline we impose upon ourselves, a brake applied to the impulsive urges of our feelings. A man who follows this arduous path will discover that it is rare in a controversy to find that one party is completely right and the other completely wrong. He will sense the futility of the reasons commonly invoked to launch one nation into war against other. In this respect, it is reasonable to say that neutrality constitutes a first step towards peace. (Pictet 1979)

Pictet employed hyperbole and caricature to reinforce the idea of political neutrality. He wrote: "Certain schools of thought maintain that everything, in the existence of a nation or even of an individual, is subordinate to political or ideological requirements." Hence, he wrote, The Red Cross "must resist this trend with every ounce of energy. Once it entered this field in which powerful forces are running wild, the Red Cross would be creating dissension within its own ranks; it would be divided against itself and head towards inevitable disaster" (Pictet 1979).

3 Medical Neutrality and Political Engagement: Pictet's Complex Legacy

The ICRC's new fundamental principles proved highly influential in the humanitarian community. In 1991, the UN General Assembly adopted guiding principles for humanitarian aid that contained the principle of neutrality. (UN General Assembly 1991). (The principle of independence was added in 2004). These principles are now accepted as providing "the foundation for humanitarian action" and claimed by the UN Office for the Coordination of Humanitarian Affairs to be essential to effective humanitarian coordination and response (UN OCHA 2012). Neutrality is often considered important for humanitarian organizations in gaining access to populations in need, negotiating with combatants, and increasing the security of humanitarian operations.

In his commentary on the ICRC's fundamental principles, Pictet did not address how his conception of humanitarian neutrality applied to the vast majority of health workers practicing in conflicts and politically volatile settings who are not affiliated with humanitarian organizations. But both scholars and the ICRC itself have assumed that all health workers are bound to be politically neutral (ICRC 2014; Mehring 2014), and the phrase medical neutrality implies as much.

At the same time, skepticism about or pushback against the demand for neutrality among leading humanitarian organizations emerged shortly after the ICRC's adoption of the new fundamental principles. And paradoxically, in some circumstances, the image of health workers as politically neutral became an enabler of political speech and action. Pictet's legacy for the conduct of health workers became even more complicated as medical ethics evolved to include obligations to promote population health and call out human rights violations.

3.1 Reinterpretations and Critiques of Political Neutrality

In 1971, Nigeria's Biafra state sought to secede from the country. The government responded with brutal repression, violence, and a blockade that led to the mass starvation of Biafrans. In light of its governing principle of neutrality, the ICRC was reluctant to speak out against the atrocities it witnessed that brought so much suffering. A group of French doctors, who were both committed to a humanitarian response and objected to the ICRC's stance, founded a new organization, Médecins Sans Frontières (MSF). As Peter Redfield explains, MSF sought to be a "more engaged and daring version of the Red Cross" (Redfield 2013).

MSF reinterpreted humanitarian neutrality, emphasizing that to achieve its commitment to relieve the suffering of the people it served it must raise its voice about the horrors inflicted on them. It adopted the idea of *témoignage*, or witnessing, which became a central value of its work and stance, In doing so, it still characterized itself as neutral, grounded in humanitarian obligations to serve affected communities but also employing medical expertise to assess the sources of ill-health. Through these two characteristics, as Redfield put it, MSF could be "both universal and restricted, impassioned and removed," with the space, authority, and a mandate to be political even as it asserted its neutrality (Redfield 2013). Recent examples of this approach are evident in detailed public statements, reports and analysis of assaults on hospitals where it worked in the Central African Republic, Gaza, Syria, and Yemen and their impacts on populations it serves. MSF's decisions whether to speak out and how to frame its messages have been context-dependent, and sometimes inconsistent, but always assess the consequences of speaking to the people it serves (including the risk of being forced out of the country) and the potential impact of its voice.

Today, humanitarian organizations interpret neutrality along a wide spectrum, from following the strict political neutrality of the ICRC to outspokenness on human rights violations. Hugo Slim, a major humanitarian thinker and former policy director at ICRC, argues for dispensing with the humanitarian principle of neutrality altogether. He notes that out that nothing in international humanitarian law requires humanitarian neutrality. More pointedly, he argues that political neutrality is inconsistent with humanitarian obligations of humanity and impartiality. These moral duties, he says, require aid groups to counter injustice, challenge tyranny, and defend the rights and lives of the people humanitarians serve. By contrast, he writes, neutrality is "myopic in its narrow focus on suffering alone and its blinkered abstention from engagement on wider political ethics." (Slim 2022). Slim goes even further, urging that, given the pervasive human rights violations in contemporary wars, aid groups adopt a stance of what he calls humanitarian resistance, toward "rescue, relief and protection of people suffering under an unjust enemy regime." (Slim 2022). He argues that the means of resistance can include demonstration, disobedience, strikes, boycotts, and sit-ins as well as speaking out, reporting, and organizing. Slim's concept humanitarian resistance is likely a bridge too far for many humanitarian organizations, but challenges Pictet's premise that political speech and action is inconsistent with humanitarian values.

3.2 *Images of Neutrality as Amplification of Medical Voices*

The principle of neutrality and, as important, image of health workers as neutral has, in some circumstances, paradoxically provided space for medical voices where they otherwise might not be heard. In 2008, Dr. Izzeldin Abuelaish, a physician living in Gaza, was practicing in a large public hospital in Israel, returning to his home on weekends. He viewed the hospital as a haven, where the staff was dedicated to equal treatment for all patients and operated entirely outside of politics. This corresponded to the perspective of Israeli doctors, who viewed medical care in Israel as apolitical, even in the content of ongoing conflict with Palestinians (Orr et al. 2022). In December of that year, Israel launched operation Cast Lead against Hamas in Gaza, starting with air and missile strikes. In January 2009 it launched a ground invasion. During the conflict, almost 1400 Palestinians were killed, 4000 houses were destroyed, multiple hospitals hit, and ambulances prevented from evacuating the wounded and often shot at. (Rubenstein 2021). Israeli media reported little about the extent of death and injury to Palestinians (they reported on Hamas rockets fired into Israel, which killed three civilians). One reason for the gap was the lack of access of Israeli media to Gaza; but another was their tradition of paying little attention to reports from Palestinians, which they deemed not credible. (Shalev 2016).

In one terrible incident during Cast Lead, Israeli tanks fired at Dr. Abuelaish's house, destroying it and killing three of his children and a niece. Distraught, he called into an Israeli TV station, and spoke to a reporter during a live broadcast. The recording of the anguished four-minute exchange was so compelling and had the credibility of a health workers that it went viral in Israel. Dr. Abuelaish subsequently published a memoir, *I Shall Not Hate: A Gaza doctor's journey on the road to peace and human dignity,* in Canada, where he immigrated with his surviving children in the aftermath of the attack. A Hebrew translation was published in Israel in March 2011 (Shalev 2016).

Anthropologist Guy Shalev wrote about Dr. Abuelaish's case, both the suffering of his family and the extraordinary success he had in communicating the violence inflicted by Israeli forces to an Israel audience that usually looked away from horrors Palestinians experience in Gaza. Dr. Abuelaish recounted his personal tragedies as a Palestinian under occupation while simultaneously invoking the social authority as a neutral physician. (Shalev 2016). Shalev argues that Dr. Abuelaish's ability to communicate suffering from his medical perspective, and one who worked side by side with Israeli doctors, gave him credibility and enabled him to be heard by Israelis on the fraught political issues of the war.

Though the circumstances of MSF *témoignage* and Dr. Abuelaish's anguished speech were vastly different, both rely on respect medical authority and apparent distance from politics even as they convey a political message. There are limitations, however, to relying on medical authority in political matters. As Shalev puts it, "Abuelaish's reliance on medical professionalism as a vehicle for communicating Palestinian grievances to outside audiences bears the risk that the public it constructs might recognize the human through the doctor, but not the nationalist claims of the Palestinian." (Shalev 2016). Moreover, the power of reliance of neutrality by health professionals to communicate effectively on the experience of war remains fragile, as Israeli dismissals of accounts by Palestinian physicians in the Gaza in the 2023–24 war illustrate.

4 Political Neutrality and Contemporary Medical Ethics

The atrocities of World War II led to revisions of the Geneva Conventions that expanded protections for civilians and health care in war and stimulated new thinking about medical ethics. The judgment of the Nuremburg war crimes tribunal in the trial of doctors involved in horrific medical experiments in concentration camps set out principles of protection of human subjects in medical research, which came to be known as the Nuremburg Code. In the same period, the World Medical Association (WMA) was established to provide global leadership on medical ethics. The WMA adopted the Declaration of Geneva, a modern version of the Hippocratic Oath, and issued an International Code of Medical Ethics. These developments in professional ethics and those that followed have had significant implications for human rights reporting and political activism by health professionals in situations of armed conflict and political volatility.

4.1 Social Obligations of Health Professionals

Despite the post-war blossoming of medical ethics, it took decades before discussion about the social responsibilities of physicians and other health professionals began. The concept that doctors have obligations beyond those to individual patients in clinical settings was not new. Rudolf Virchow, the nineteenth century pioneer of social medicine, wrote that the physician is a "natural attorney for the poor," with responsibilities to improve population health, especially for the marginalized and disenfranchised (Brown and Fee 2006). In the post-World War II period, however, organized medicine remained conservative, especially in developed countries, and deeply enmeshed in the colonialist and racist attitudes of their societies. Until this century, professional ethical codes remained principally concerned with ethics in clinical practice and often tolerated racial and other forms of discrimination. And despite the adoption of the Universal Declaration of Human Rights in 1948, the WMA's Declaration of Geneva and International Code of Medical Ethics were silent on physician obligations on human rights and population health.

In the last two decades of the twentieth century, however, views of the social obligations of the health professionals began to change, likely influenced by a new medical activism. In 1985, International Physicians for the Prevention of Nuclear War, founded in 1980, was awarded the Nobel Peace Prize. Medical groups in South Africa and globally became active the anti-Apartheid movement. Amnesty International took up human rights violations by and against health workers, and new human rights organizations focusing on health concerns documented abuses and pressed for reforms.

By the turn of the new century, health professional associations began to recognize social obligations as part of practitioners' ethical responsibilities. The WMA's Declaration of Geneva was amended to state that "I will share my medical knowledge for the benefit of the patient and the advancement of healthcare." (WMA 2017). Its International Code of Medical Ethics was revised to oblige physicians to address inequities in health and care, the determinants of those inequities, and violations of the rights of patients and health professionals. It recognized an obligation to "contribute to the health and well-being of the populations the physician serves and society as a whole, including future generations." (WMA 2022). The International Council of Nurses similarly required nurses to advocate for equity and social justice and identified specific forms of exploitation that nurses must oppose, including human trafficking and child labor. It requires nurses to advocate for policies and programs that address social determinants of health (International Council of Nurses 2021).

The WMA also established more specific requirements on human rights reporting by physicians. Its Declaration of Tokyo, originally adopted in 1975, forbade any form of physician participation in torture and other forms of cruel and inhuman treatment. In this century, the WMA expanded the obligation to notifying relevant authorities of evidence of torture they found in medical records or in clinical or forensic examinations. Doing so, it said, so can contribute "to the human rights of the victims and to the protection of their physical and mental integrity." (WMA 2016). The WMA also offered guidance on addressing situations where reporting obligations might conflict with other ethical responsibilities, such as obtaining informed consent and maintaining patient confidences. It took an approach consistent with medical ethics generally, recognizing possibly competing values and reaching a decision by balancing the benefits, burdens, and risks of different courses of action.

Its approach to the obligation to report torture is illustrative. The WMA recognized that individuals subjected to torture could be subject to retaliation and even further torture if their abuse is reported. It therefore counseled physicians to "carefully balance potential conflicts between their ethical obligation to document and denounce acts of torture or cruel, inhuman and degrading treatment and punishments and a patient's right to informed consent before documenting torture cases." Consistent with approaches to resolving dilemmas regarding conflicting obligations—here, the importance of reporting versus gaining consent and maintaining confidentiality—it stated that patient consent is not necessarily determinative. Rather, the guidelines of the Declaration of Tokyo now state that "physicians have the ethical obligation to report abuses, where possible with the subject's consent, but in certain circumstances where the victim is unable to express him/herself freely, without explicit consent." (WMA 2016).

It was this balancing approach the Myanmar doctors employed in their explanation of the reasons for their participation in their protests and strikes. They recognized the ethical complexity of the decision and forthrightly recognized the potential harms to their patients from their actions. But they also considered the deterioration of the nation's health care system and COVID-19 response after the coup and the impossibility of providing ethical and quality care to the population under the thumb of a military dictatorship. Balancing the potential harms of their departure from public hospitals and the need to resist an oppressive and cruel military government, they chose to strike.

4.2 *ICRC Ethical Guidance and the Legacy of Pictet*

In 2014, the ICRC published a guidebook on the responsibilities of health workers in armed conflict or situations of state repression, civil unrest, and demonstrations. It was explicitly designed for all health workers, whether they practice in a humanitarian organization, government, private practice, or an NGO health facility. The guidance embraces the now-accepted duty to advocate for impartial care and to "remind authorities of their obligations," and, if possible, "to give consideration to how you and other health-care personnel might curtail or mitigate the consequences of the conflict or emergency in question by, for example, tackling violations of humanitarian law or human rights law." (ICRC 2014).

The guidance also recognizes the dilemmas that health workers who witness human rights violations face. It notes that they may "feel a pressing need to pass on such information to a competent authority or institution or even the media," with the hope of stopping the violation. But it warns that reporting could create risks of harm to the health worker, the organization, or the people served. The guidance purports to follow the traditional process of resolving ethical dilemmas by balancing risks and benefits in pursuing a course of action given the conflicting values at stake.

The guidance is expressly premised, however, on the idea of the "neutrality of all health care." It does not explain the source of this assertion, but likely refers to the humanitarian principle of neutrality, which severely limits political speech and action. The guidance's more specific instructions, moreover, are highly constraining in their view of the ethical propriety of collecting data, reporting on, and speaking out about human rights abuses. It puts a thumb on the scale in balancing competing obligations toward refraining from speech or action. Its advice is replete with draconian conditions on data collection and reporting and stern warnings about possible consequences that discourage health professionals from collecting information about human rights violations, reporting them and, especially, speaking with the media about them.

For example, in its discussion of gathering data about possible human rights violations, the guidance advises that where the health worker is "concerned" that publication "might" endanger the health worker, the wounded or sick, or colleagues, they should limit data collection to routine health information contained in existing records; in other words, they should not collect information on human rights violations at all. Given the commonality of "concerns" that such risks "might" occur, the guidelines can only be read to collect data only in the most anodyne situations rather than balancing risks and benefits of doing so or considering ways to mitigate them. Its discussion, moreover, forgoes any consideration comparable to the WMA's approach of balancing security concerns against the need to collect and report data that could prevent future harm to individuals and the populations. And it counsels that in the most difficult and precarious situations not to collect data at all; it contains no discussion how to collect data safely in such circumstances. If followed, this approach would mean not collecting information even in cases where fact-gathering is most essential for protection of patients and the civilian population more broadly, and where rational, ethical health workers can decide to take such risks, especially to themselves, after thoughtful balancing of risks and benefits. As the guidance takes such a strong stance of avoiding the possibility of risks, it says nothing about efforts to mitigate them.

The guidance takes a similarly cramped stance on reporting. It expresses a strong preference for publication in professional journals or at legal or diplomatic conferences, even at is recognizes that the impact of this course of action may only be "in the long run," not in real time. It ignores both the growth and impact of human rights reporting by medical groups and individual health workers in armed conflict and evidence that authors of these reports have not compromised anyone's safety. Again, too, the guidelines eschew balancing of benefits, burdens, and risks altogether, instead suggesting health professionals constrain how they report. This approach is evident, too, in the guidance's discussion of media. It urges health professionals not to speak to the media about data collection unless the health worker is "absolutely sure" there is no danger to you or others, an impossible standard to meet. It adds other categorical admonitions, for example that privacy of the patient always takes precedence over what it characterizes as the "media's curiosity or needs." This approach is directly contrary to that of the WMA's advice on reporting torture. And it does not recognize that the media can be essential in raising public awareness and understanding of severe human rights violations and stimulating action to stop them.

Pictet's influence seems to haunt the guidance, both in its opposition to engagement in political acts and emotionally charged warnings about breaches of professional duties. In its discussion of speaking to the media, the guidance warns that if a health worker takes advantage of their position and "your supposed credibility as a witness" regarding violations of law, "you endanger not only your own safety and that of others, but you also run the risk of escalating the process of accusation and counter-accusation that is a feature of any armed conflict or other emergency. This is serious professional misconduct." Such shrill hyperbole has no place in serious discussions of how to address human rights violations. Verification is often difficult in war zones, and it is often important to be a witness and make judgments, as when unconfirmed information can trigger investigations by others in a position to verify. That a perpetrator may "escalate" by denying the allegation or seeking to discredit the reporter is to be expected and not a reason to refrain from reporting. Neither does raising issues in good faith based on sources deemed by a health professional to be credible, even if not verified; stimulating a negative response is far from professional misconduct. Even the phrasing of "accusations and counter-accusations" is reminiscent of Pictet, as though a bad faith denial by perpetrators of abuse is morally equivalent to good faith reporting.

It is striking that the ICRC's guidance ignores the emergence of health professionals' ethical obligations toward protection of human rights and health equity over the past generation. It is equally concerning in paying scant heed to the costs of not collecting data or reporting, especially the risks of continuing human rights violations and deprivations of health care if abuses are not documented and reported. In addition, it fails to address the potential reputational costs to organizations and health professionals if they depart from contemporary ethical standards regarding political engagement and instead follow the stern advice the ICRC urges.

For example, the Burmese military, the Tatmadaw, long committed atrocities against civilians during an armed conflict in eastern Myanmar where ethnic groups sought autonomy from the central government. Its assaults displaced many thousands of people, leaving them without access to health care. During the 1990s, indigenous organizations sought to fill the gap. One of them, the Back Pack Health Workers Team, trained and dispatched community health workers to the most insecure and at risk displaced populations. In doing so, the health workers themselves became vulnerable to attack and kidnapping by the Tatmadaw (Footer et al. 2014). Back Pack's teams were widely admired for their professionalism and effectiveness and the courage its health workers. But because the group also has had an explicitly pro-democracy agenda, and close ties with indigenous groups with armed affiliates, its international donors were sometimes reluctant to provide funding it because its lack of political neutrality. (Decobert 2015). Most recently, in Ukraine, local health NGOs have faced similar skepticism by donors for lack of neutrality in the war in speaking about Russia's aggression and war crimes (personal communication).

Another question arises as well: Do demands for political neutrality themselves place health workers at risk of attack? We lack evidence whether the assaults on health workers in Sudan, Turkey, Myanmar, Syria and elsewhere have been rationalized as justified because of their lack of political neutrality when supporting political reform or speaking out. It is a serious question, though, and there are indications that political engagement by health professionals has been used as an excuse to harm them. In Syria, detained doctors were sometimes singled out for especially brutal treatment, including torture (Rubenstein 2021).

5 Conclusion

Obligations of human rights data collection and reporting, political speech, and activism are now firmly established in health professional ethics and consistent with the Geneva Conventions and moral obligations of humanity and impartiality. Their voices can be critical to the well-being, dignity, and protection of the individuals and populations and they serve. Suggesting instead that all health workers, not just those affiliated with humanitarian organizations, must instead remain mostly silent in the face of atrocities, oppression by tyrannical regimes, and assaults on health care itself, is misguided, contrary to contemporary ethics and, in some instances, dangerous. That stance suppresses knowledge that only they may be able to gather, stills potentially powerful voices, and in the worst cases may allow breaches of fundamental rights of people at great risk to continue. Instead, as in all other aspects of ethical health practice, engaging in political acts must be undertaken thoughtfully, including by weighing benefits and risks of different courses of action and possibilities for mitigating harms in reconciling possibly conflicting obligations to individual patients and populations.

It is also time to discard the phrase "medical neutrality" altogether. Its multiple meanings serve only to confuse, and to the extent it conveys demands for political neutrality, it eschews context dependent and considered judgments consistent with ethical principles. Pictet's ghost should no longer haunt.

References

Abuelaish, Izzeldin. 2011. *I shall not hate: A Gaza doctor's journey on the road to peace and human dignity*. New York: Walker & Company.

Brown, Theodore, and Elizabeth Fee. 2006. Rudolf Carl Virchow: Medical scientist, social reformer, role model. *American Journal of Public Health* 96 (12): 2104–2105.

Convention (I) for the amelioration of the condition of the wounded and sick in armed forces in the field. Geneva, 12 August 1949. https://ihl-databases.icrc.org/en/ihl-treaties/gci-1949. Accessed 24 Aug 2023.

Convention for the amelioration of the condition of the wounded and sick in armies in the field. Geneva, 6 July 1906. https://ihl-databases.icrc.org/en/ihl-treaties/gc-1906. Accessed 24 Aug 2023.

Convention for the amelioration of the condition of the wounded in armies in the field. Geneva, 22 August 1864. https://ihl-databases.icrc.org/en/ihl-treaties/gc-1864. Accessed 24 Aug 2023.

Decobert, Anne. 2015. *The politics of aid in Myanmar: A humanitarian struggle on the Thai-Burmese border*. London: Routledge.

DLP Forum. 2023. *The law of neutrality*. https://www.dlpforum.org/2023/05/05/the-law-of-neutrality/ Accessed 24 Aug 2023.

Dunant, Henri. 1959. *A memory of Solferino*. Geneva: ICRC.

Footer, Katherine, Sarah Meyer, Susan Sherman, and Leonard Rubenstein. 2014. On the frontline of eastern Burma's chronic conflict: Listening to the voices of local health workers. *Social Science & Medicine* 120: 378–386.

Huber, Max. 1948. The principles of the red cross. *Foreign Affairs* 26 (4): 723–727.

ICRC. 2014. *The responsibilities of health-care personnel working in armed conflicts and other emergencies*. https://healthcareindanger.org/wp-content/uploads/2015/09/icrc-002-4104-the-responsibilities-health-care-personnel.pdf. Accessed 24 Aug 2023.

———. 2016. *Fundamental principles of the red cross and red crescent movement*. https://www.icrc.org/en/document/fundamental-principles-red-cross-and-red-crescent. Accessed 24 August 2023.

ICRC and International Federation of Red Cross and Red Crescent Societies. 2020. *The fundamental principles of international red cross and red crescent movement: Ethics and tools for humanitarian action*. Geneva: ICRC.

International Council of Nurses. 2021. *ICN code of ethics*. https://www.icn.ch/system/files/documents/2020-10/CoE_Version%20for%20Consultation_October%202020_EN.pdf. Accessed 24 Aug 2023.

Kalshoven, Frits. 1992. International humanitarian law and violation of medical neutrality. In *Violation of medical neutrality*, ed. G.L. Wackers and C.T.M. Wennekes. Amsterdam: Thesis Publishers.

Kenny, Carolina. 2015. Emmerich de Vattel, The Law of Nations (1758). Classics of Strategy and Diplomacy. https://classicsofstrategy.com/2015/08/07/law-of-nations-vattel-1758/ Accessed 24 Aug 2023.

Kolb, Robert, and Fumiko Nakashima. 2019. The notion of "acts harmful to the enemy" under international humanitarian law. *International Review of the Red Cross* 1019: 1171–1199.

Labbe, Jeremie, and Pascal Daudin. 2016. Applying the humanitarian principles: Reflecting on the experience of the International Committee of the Red Cross. *International Review of the Red Cross* 97: 183–210.

League of Red Cross Societies. Board of Governors, XIXth session, Oxford, 1946, Resolution 12.

Mehring, Sigrid. 2014. *First do no harm: Medical ethics in international humanitarian law*. Leiden: Brill Neijhoff.

Orr, Zvika, Albert Jackson, Evan Avraham, and Mark Fleming. 2022. Neutrality, conflict, and structural determinants of health in a Jerusalem emergency department. *International Journal for Equity in Health* 21: article 89.

Pictet, Jean. 1979. *The fundamental principles of the Red Cross*. Geneva: International Federation of Red Cross and Red Crescent Societies.

Protocols I and II additional to the Geneva Conventions. 1977. https://www.icrc.org/en/doc/assets/files/other/icrc_002_0321.pdf. Accessed 24 Aug 2023.

Redfield, Peter. 2013. *Life in crisis: The ethical journey of doctors without borders*. Berkeley: University of California Press.

Rubenstein, Leonard. 2021. *Perilous medicine: The struggle to protect health care from the violence of war*. New York: Columbia University Press.

Shalev, Guy. 2016. A doctor's testimony: Medical neutrality and the visibility of Palestinian grievances in the Jewish-Israeli publics. *Culture, Medicine, and Psychiatry* 40: 242–262.

Slim, Hugo. 2022. *Humanitarian resistance: Its ethical and operational importance*, 87. London: Humanitarian Practice Network. https://odihpn.org/wp-content/uploads/2022/09/Humanitarian-resistance_NP_web.pdf. Accessed 24 Aug 2023.

Soe, Zaw Wei, Maw Maw Oo, Shwe Wah Khine, Aye Thiri Naing, Rosanne Skalicky-Klein, and Georgina Phillips. 2021. Myanmar's health leaders stand against military rule. *The Lancet* 397: P875.

Steinacher, Gerald. 2017. *Humanitarians at war: The red cross in the shadow of the holocaust*. Oxford: Oxford University Press.

U.N General Assembly. 1991. Resolution 46/182.
U.N. General Assembly. 2013. Report of the special rapporteur on the highest attainable standard of physical and mental health, A/68/297.
U.N. Office for the Coordination of Humanitarian Activities. 2012. *OCHA on message: Humanitarian principles.* https://time.com/wp-content/uploads/2015/02/oom-humanitarianprinciples_eng_june12.pdf. Accessed 24 Aug 2023.
U.S. Department of Defense. 2023. *Law of war manual.*
World Medical Association. 2011. *Physicians must be considered as neutral in armed conflicts and civil unrest.* https://www.wma.net/news-post/physicians-must-be-considered-as-neutral-in-armed-conflicts-and-civil-unrest/. Accessed 24 Aug 2023.
———. 2016. WMA Declaration of Tokyo—Guidelines for physicians concerning torture and other cruel, inhuman or degrading treatment or punishment in relation to detention and imprisonment. https://www.wma.net/policies-post/wma-declaration-of-tokyo-guidelines-for-physicians-concerning-torture-and-other-cruel-inhuman-or-degrading-treatment-or-punishment-in-relation-to-detention-and-imprisonment/#:~:text=The%20physician%20shall%20not%20countenance,and%20in%20all%20situations%2C%20including. Accessed 24 Aug 2023.
——— 2017. *Declaration of Geneva.* https://www.wma.net/policies-post/wma-declaration-of-geneva/ Accessed 24 August 2023.
———. 2022. *International code of medical ethics.* https://www.wma.net/policies-post/wma-international-code-of-medical-ethics/. Accessed 24 Aug 2023.

The manufacturer's authorised representative in the EU is Springer Nature Customer Service Centre GmbH, Europaplatz 3, 69115 Heidelberg, Germany. If you have any concerns regarding our products, please contact ProductSafety@springernature.com

Printed and bound by CPI Group (UK) Ltd, Croydon, CR0 4YY
30/09/2026
02225564-0002